Ordered Media
in Chemical Separations

ACS SYMPOSIUM SERIES **342**

Ordered Media in Chemical Separations

Willie L. Hinze, EDITOR
Wake Forest University

Daniel W. Armstrong, EDITOR
Texas Tech University

Developed from a symposium sponsored by
the Division of Analytical Chemistry
at the 191st Meeting
of the American Chemical Society,
New York, New York,
April 13–18, 1986

American Chemical Society, Washington, DC. 1987

Library of Congress Cataloging-in-Publication Data

Ordered media in chemical separations.
 (ACS symposium series; 342)

 Developed from the Symposium on Use of Ordered
Media in Chemical Separations.

 Includes bibliographies and indexes.

 1. Chromatographic analysis—Congresses.
2. Separation (Technology)—Congresses. 3. Surface
active agents—Congresses. 4. Cyclodextrins—
Congresses.

 I. Hinze, Willie L., 1949-   . II. Armstrong, Daniel
W., 1949-   . III. American Chemical Society.
Division of Analytical Chemistry. IV. American
Chemical Society. Meeting (191st: 1986: New York,
N.Y.) V. Symposium on Use of Ordered Media in
Chemical Separations (1986: New York, N.Y.)
VI. Series.

QD79.C4073   1987   543'.08   87-13563
ISBN 0-8412-1402-6

*QD
79
.C4
Q73
1987*

Foreword

The ACS SYMPOSIUM SERIES was founded in 1974 to provide a medium for publishing symposia quickly in book form. The format of the Series parallels that of the continuing ADVANCES IN CHEMISTRY SERIES except that, in order to save time, the papers are not typeset but are reproduced as they are submitted by the authors in camera-ready form. Papers are reviewed under the supervision of the Editors with the assistance of the Series Advisory Board and are selected to maintain the integrity of the symposia; however, verbatim reproductions of previously published papers are not accepted. Both reviews and reports of research are acceptable, because symposia may embrace both types of presentation.

Contents

Preface

ENORMOUS ADVANCES AND GROWTH IN THE USE OF ORDERED MEDIA (that is, surfactant normal and reversed micelles, surfactant vesicles, and cyclodextrins) have occurred in the past decade, particularly in their chromatographic applications. New techniques developed in this field include micellar liquid chromatography, micellar-enhanced ultrafiltration, micellar electrokinetic capillary chromatography, and extraction of bioproducts with reversed micelles; techniques previously developed include cyclodextrins as stationary and mobile-phase components in chromatography. The symposium upon which this book was based was the first major symposium devoted to this topic and was organized to present the current state of the art in this rapidly expanding field.

This volume resulted from the need to have a readily available reference source to present an account of the roles and uses of ordered media in separation science. The emphasis has been placed on chromatographic applications. The organization of the volume divided naturally into two parts. The first part, consisting of 10 chapters, deals with organized surfactant media in separation science; the second, consisting of six chapters, details the applications of cyclodextrins in chromatography. Emphasis has been placed on a critical assessment of recent work and an integration of material from a wide range of sources. Although all of the applications involving these types of ordered media in separation science were not covered, it is hoped that all of the important recent developments in this field have been included.

We thank the authors for their contributions and their interest in this project. We are also grateful to the anonymous referees for their time, invaluable comments, and constructive criticism of the manuscripts during the review process. We also thank the ACS Books Department staff for their help with this undertaking and for their patience.

WILLIE L. HINZE
Wake Forest University
Winston–Salem, NC 27109

DANIEL W. ARMSTRONG
Texas Tech University
Lubbock, TX 79409

SURFACTANT AND RELATED SYSTEMS

Chapter 1

Organized Surfactant Assemblies in Separation Science

Willie L. Hinze

Department of Chemistry, Wake Forest University, P.O. Box 7486, Winston-Salem, NC 27109

A brief description of the structural features and relevant properties of different organized assemblies formed from surfactant molecules is presented. Next, the use and application of these organized surfactant systems in separation science is surveyed. Several possible new areas for future developments employing these ordered media are mentioned.

Many separation processes mediated by the presence of surfactant organized assemblies (also referred to as organized or ordered media) have been developed during the past ten years. The growing importance and popularity of such separation techniques is demonstrated by the fact that numerous recent review articles have been devoted to this subject (1-10). The purpose of this overview is two-fold. First, it is intended to provide the novice entering the field with basic, simplified background information on organized surfactant systems which should facilitate a better understanding of the more specific technical articles that appear in subsequent chapters of this monograph (or the chemical literature). Secondly, this overview will attempt to update and summarize the previous reported work in this area of separation science. Topics not extensively covered in the previous reviews (or succeeding chapters of this monograph) will be discussed in greater detail. Throughout, emphasis will be placed on the practical applications and potential future developments. It is hoped that this overview will paint a general picture of the structure, properties, and role of different surfactant organized assemblies in separation science.

Structure and Properties of Different Organized Surfactant Assemblies

Structure Formation in Surfactant Solutions. Surfactants, also referred to as soaps, detergents, tensides, or surface active agents, are amphiphilic molecules possessing both hydrophilic and hydrophobic regions. They can be classified as anionic, cationic, zwitterionic, or nonionic (neutral) depending upon the nature of the polar

0097-6156/87/0342-0002$18.30/0

head-group that is bound to the nonpolar hydrocarbon tail. Various
colloidal-sized organized structures can form when surfactant
molecules are dissolved in a particular solvent depending upon the
nature and concentration of the surfactant molecule, nature of the
solvent system, and exact experimental conditions (i.e. temperature,
pressure, and/or presence or absence of additives) (11-16). Figure 1
shows an oversimplified representation of some of the different
surfactant species possible as the surfactant concentration is
increased in a surfactant – water two-component system. At low
concentrations above the Krafft temperature, the surfactant is present
in isolated monomeric molecular form. With further increases in
concentration, the surfactant molecules can dynamically associate to
form micellar assemblies (termed aqueous or normal micelles). The
surfactant concentration at which such aggregation occurs is referred
to as the critical micelle concentration (CMC) and the number of
surfactant molecules comprising the micellar entity is called its
aggregation number (N). Although aqueous normal micelles are
generally viewed as being roughly spherical (Figure 1), considerable
controversy still exists concerning the exact shape and structure of
such entities (15,17). Typically, such micellar aggregates are
composed of 40 – 140 monomeric surfactant molecules such that their
hydrophobic tails are oriented inward forming a nonpolar core region
and their hydrophilic headgroups are directed toward and in contact
with the bulk aqueous solvent. Further increases in surfactant
concentration can result in the formation of other different types of
organized assemblies. Initially, there can be a transition from
spherical to rodlike or cylindrical micelles (Figure 1). Still higher
concentrations lead to formation of various liquid crystalline
aggregates (Figure 1:   middle, viscous, and neat liquid crystalline
phases) (11,15,18). The presence of a third component (organic
solvent) can give rise to an even larger variety of aggregated
surfactant species (15). Table I summarizes the structure, name, and
micellar parameters (CMC and N) of some typical long-chain alkyl
surfactants employed to form aqueous normal micellar systems (11,19).

   In addition to these types of micellar-forming surfactants, there
is another class of molecules that can associate in water to form
micellar aggregates; namely, the bile salts (20). Bile salts are very
important biological detergent-like molecules. However, they differ
from the long-chain alkyl surfactants previously mentioned in that
they possess a hydrophobic and a hydrophilic face (Figure 2).
Consequently, bile salts exhibit a different type of aggregation
behavior. That is, the aggregation process is viewed as consisting of
the stepwise formation of initial primary micelles which are composed
of 2 – 8 monomers held together by hydrophobic interactions between
the bile salt nonpolar faces. At higher bile salt concentration (or
high ionic strength), the primary micelles can further aggregate to
form larger, rod-like cylindrically shaped secondary bile salt
micelles due to intermolecular hydrogen bonding between their hydroxyl
groups (2,21). Table II presents the name, structure, and micellar
parameters of some common bile salts. All surfactants and bile salts
mentioned in Tables I and II are commercially available (22). A
recent multi-volume series lists the trade name, chemical name,
manufacturer, form, properties, toxicity, composition, principal and
secondary uses, etc. for many of these surfactants (26).

TABLE I. Structure, Name, Abbreviation, and Micellar Parameters of
          Some Aqueous Micellar-Forming Surfactants Employed in
          Separation Science

Surfactant Structure; Name; and [Abbreviation]	CMC,[a,b] mM	N[b,c]
Anionic Micelle-Forming Surfactants of General Formula $R-X^- M^+$:		
$R=C_{16}$, $X=OSO_3$, M=Na; Sodium hexadecylsulfate [NaHDS]	0.52	100
$R=C_{12}$, $X=OSO_3$, M=Na; Sodium dodecylsulfate [NaLS]	8.1	62
$R=C_{10}$, $X=OSO_3$, M=Na; Sodium decylsulfate [NaDS]	33.0	50
$R=C_8$, $X=OSO_3$, M=Na; Sodium octylsulfate [NaOS]	136.0	20
$R=C_{10}$, $X=CO_2$, M=Na; Sodium laurate [NaL]	24.0	56
$R=C_{10}$, $X=CO_2$, M=K; Potassium laurate [KL]	12.5	48
$R=C_{15}$, $X=CO_2$, M=K; Potassium palmitate [KP]	4.0	--
$R=C_{12}$, $X=SO_3$, M=Na; Sodium dodecylsulfonate [NaDDS]	9.8	54
Cationic Micelle-Forming Surfactants of General Formula $R-N^+(CH_3)_3 X^-$:		
$R=C_{16}$, X=Cl; Hexadecyltrimethylammonium chloride [CTAC]	1.3	78
$R=C_{16}$, X=Br; Hexadecyltrimethylammonium bromide [CTAB]	0.9	61
$R=C_{12}$, X=Br; Dodecyltrimethylammonium bromide [LTAB]	15.0	50
$R=C_{10}$, X=Br; Decyltrimethylammonium bromide [DTAB]	65.0	47
$R=C_8$, X=Br; Octyltrimethylammonium bromide [OTAB]	180	--
Mixture of $R=C_{12}$, C_{14}, and C_{16}, X=Br; Cetrimide [C] predominately $R=C_{14}$	2.0-8.5	50-62
$CH_3(CH_2)_{15}N^+C_5H_5$ Cl^-; Cetylpyridinium chloride [CP]	0.9	95
$CH_3(CH_2)_{15}N^+(CH_3)_2(CH_2C_6H_5)Cl^-$; Hexadecyldimethyl-benzylammonium chloride [CBzAC]	0.27	--
Nonionic Micelle-Forming Surfactants of General Formula $R(OCH_2CH_2)_nOH$:		
$R=C_{12}$, n=23; Polyoxyethylene(23)dodecanol [Brij-35]	0.1	40
$R=(CH_3)_3CCH_2C(CH_3)_2C_6H_4$, n=9.5; Polyoxyethylene-t-octylphenol [Triton X-100 or TX-100]	0.2	143
$R=CH_3(CH_2)_7CH=CH(CH_2)_8$, n=10; 10 Oleyl ether [Brij-96]	0.04	--
R=mixture of C_9, C_{10}, and C_{11}; n=6; Neodol 91-6	0.37(wt%)-	
Polysorbate 80 (or Polyoxyethylene sorbitan mono-oleate [Tween-80]	0.012	60
Zwitterionic Micelle-Forming Surfactants of General Formula $R-(CH_3)_2N^+CH_2X^-$:		
$R=C_{10}$, $X=CH_2CH_2SO_3$; N-Decylsultaine [SB-10]	3.9	--

Surfactant Structure; Name; and [Abbreviation]	CMC,[a,b] mM	N[b,c]
$R=C_{12}$, $X=CH_2CH_2SO_3$; N-Dodecylsultaine [SB-12]	1.2	55
$R=C_{16}$, $X=CH_2CH_2SO_3$; N-Hexadecylsultaine [SB-16]	0.1	--
$R=C_{10}$, $X=CO_2$; N-Decylbetaine [DDAA]	10 - 21	34
$R=C_{12}$, $X=CO_2$; N-Dodecylbetaine [DoDAA]	1.5	73
$R=C_{16}$, $X= CO_2$; N-Hexadecylbetaine [HDAA]	0.02	--

[a]Critical micelle concentration. [b]Micellar parameters given are for aqueous solutions at 25°C, 1 atm, in the absence of any additives. Values taken from references (1,5,12,13,15,16). [c]Aggregation number (N).

molecules

spherical micelles

cylindrical micelles

middle phase

viscous isotropic phase

neat phase

Figure 1. Simplified representation of idealized surfactant species that may form in water as the surfactant concentration is progressively increased. "Reproduced with permission from Ref. 18. Copyright 1979, The Chemical Society ."

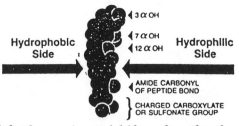

Figure 2. Model of a conjugated bile salt molecule (side view)
which shows the spatial arrangement of the hydrophobic and hydrophilic face. "Reproduced with permission from Ref. 23. Copyright 1973, Plenum Press ."

It is important to stress that the micellar parameters presented
in Tables I and II are for the indicated detergent in water at ambient
atmospheric pressure and room temperature ($\approx 25°C$). The quoted values
can be altered (sometimes dramatically) by changes in the experimental
conditions. For instance, temperature and pressure can impact the
micellization process. Typically, plots of CMC vs. temperature
exhibit a minimum somewhere between 20 - 30° C for charged ionic
surfactants while for nonionic surfactants, only a limiting minimum is
observed at ca. 40 - 50° C (16). The micellar CMC and N can also
depend upon pressure (16). However, at the pressures under which most
separation techniques are conducted ($\leq$ 3.5 MPa), the changes in
micellar parameters are such that this effect can be neglected in all
but the most exacting work (64).

More drastic changes in the CMC and N are observed when additives
are present in the micelle-forming surfactant - water systems. The
addition of ionic species (i.e. electrolytes) usually results in an
increase in the aggregation number and a reduction in the CMC. Table
III (and Table II) present some data which illustrate this effect.
Depending upon the concentration, the presence of water miscible
organic molecules can either enhance or inhibit micelle formation.
For example, short-chain alcohols can enhance micelle formation (i.e.
lower the CMC) if present at very low mole fraction and prevent
micellization at higher concentration (if X $\geq$ 0.05 or 10-15% by
volume) (27,28). Other organic solvents, like acetone, dioxane,
acetonitrile, tetrahydrofuran, etc. that form relatively strong
hydrogen bonds with water, will generally have a slight inhibitory
effect on the micellization process (i.e. greater CMC value) when
present at very low concentration (28,29). At greater concentrations
(X $\geq$ 0.10 or 15-20% by volume), their presence prevents micelle
formation. Lastly, some organic solvents (hydrazine, 1,3-propanediol,
formamide, glycerol) which can have three-dimensional structure in
their neat liquid state, can promote micelle formation if present at
relatively low concentration as well as allow for micelle formation in
mixtures of these solvents with water in all proportions (28,29). If
the organic additive is a normally water immiscible substance, then
its effect on the micellization process can be more complicated (65).
For instance, the addition of long chain alcohols (containing 5 or
more carbon atoms) or alkanes can either enhance or inhibit micelle
formation depending upon the concentration of the surfactant present

TABLE II. Structure and Micellar Parameters of Some Bile Salts[a]

Structure, Name	CMC, mM	N

if $R_1=R_2=R_3=H$; Cholanoic acid	---	--
if $R_1=R_2=R_3=OH$; Cholic acid (CA)/sodium cholate (NaC)	12.5[b]	3[c]
if $R_1=R_3=OH$; $R_2=H$; Deoxycholic acid (DCA)/sodium deoxycholate (NaDC)	6.4[b], 2.8[c]	14[c]
if $R_1=R_2=OH$, $R_3=H$; Chenodeoxycholic acid (CDCA)/sodium chenodeoxycholate (NaDCA)	5.7[b], 2.7[c]	10[c]
if the acid (position 24) is conjugated with taurine, then can have the corresponding tauro derivaties: i.e. $R_1=R_3=OH$, $R_2=H$, with position 24 acid conjugated with taurine; Taurodeoxycholic acid (TDC)/sodium taurodeoxycholate (NaTDC)	4.0[d], 1.6[c], 8.5[e]	3.5[d] 10[c]

[a]Derivatives of cholanoic acid (20). Data taken from References (20, 24, 25). [b]In 0.001 M NaOH. [c]In 0.15 M NaCl. [d]In water alone. [e]At pH 7.4.

TABLE III. Comparison of Micellar Parameters Under Different
Experimental Conditions in Aqueous Media

Surfactant	Experimental Factor Varied		CMC, mM	N
Hexadecylpyridinium Bromide (CPB)[a]	Temperature:	25°C	0.58	--
		35	0.77	--
		45	0.89	--
		55	1.0	--
	Added alcohol, Methanol:	0 (w/w) %	0.58	--
		6.4	0.75	--
		14.7	1.18	--
		19.9	1.69	--
		26.0	2.81	--
	Bulk Solvent:	Water (25°C)	0.58	--
		Ethylammonium nitrate (fused salt system, 30° C)[b]	20.0	26
Sodium Dodecyl Sulfate (NaLS)[c]	Pressure:	0.1 MPa	---	60
		40.0	---	42
		80.0	---	35
		95.0	---	38
		120.0	---	50
		140.0	---	78
	Added Electrolytes:			
		none (25°C, 1 atm)	8.1	62
		0.05 M NaOH	2.7	--
		0.10 M NaOH	1.5	--
		0.15 M NaCl	1.3	95
		0.30 M NaCl	---	117
		0.55 M NaCl	---	580
		1.0×10^{-5}M Mg^{2+}	0.7	--
		1.0×10^{-5}M Fe^{2+}	0.8	--
	Bulk Solvent:	Water (35°C)	8.57	--
		Hydrazine (35°C)[d]	22.0	--
		Formamide (60°C)[d]	220.0	--

[a]Data taken from Ref. (19) unless otherwise indicated.   [b]Taken from
Ref. (30).   [c]Data taken from Ref. (5,11,13,16,19).   [d]Taken from Ref.
(63).

and the amount of organic added. In many instances, the formation of microemulsions can result, particularly at higher alcohol or alkane concentrations (209). Consequently, the variation in micellar parameters (CMC and N) or structure with changes in the experimental conditions should be kept in mind when one uses surfactant organized assemblies in separation science applications.

Apart from micelle formation when surfactants are added to water, vesicle formation can also occur (15). Namely, if certain types of surfactants, i.e. typically long chain dialkyl-containing surfactants, are added to water and sonicated above their phase transition temperature, closed bi- or multi-layered structures called vesicles can form (15,31-36,211). Table IV lists the structure and some common properties of the most studied vesicle-forming surfactant systems. Compared to the normal micellar systems just described, such surfactant vesicles are much larger, more static (i.e. less fluid "more rigid") aggregates. Vesicle aggregates, once formed, cannot be destroyed by dilution whereas micelles can. Most synthetic surfactant vesicle systems also exhibit temperature dependent phase transition behavior in contrast to the micelle systems (15). Although the subject of much recent study, surfactant vesicles have not yet been utilized to any appreciable extent in separation science. To date, they have been employed as models for the study of biological transport and membrane - solute interactions (9). Of course, such information is useful and could lead to development of separation schemes involving surfactant vesicles - especially in the area of membrane - based separations. More information and details of such vesicular and related organized assemblies in this context is provided by the following Chapter by Fendler in this Volume (36).

In addition to structure formation in water, ordered surfactant assemblies can form in nonpolar solvents as well. For instance, when surfactant molecules are dissolved in organic hydrocarbons in the presence of small amounts of water, the formation of ion pairs as well as small and large aggregates is possible (5,8,11-16,36-41). The term reversed or inverted micelles is given to such aggregates since their polar groups are concentrated in the interior (core) region of the surfactant assembly while their hydrophobic portions extend into, and are surrounded by, the bulk nonpolar solvent molecules. The reversed micellar internal core region contains the hydrophilic headgroup of the surfactant in addition to an inner pool of co-solubilized water (or other polar solvents). It must be stressed that the usual concepts and structural models typically employed to describe normal aqueous micellar formation in water are not always applicable to reversed micellar systems in organic solvents (37-41). In fact, several modes of aggregation are possible depending upon the charge-type surfactant employed.

According to Muller's classification scheme (27,39), the majority of surfactants (i.e. cationic, zwitterionic, and most nonionic) undergo so-called Type I aggregation behavior. This is, aggregation of these surfactants proceeds via a smooth transition of monomer $\rightleftarrows$ dimer $\rightleftarrows$ trimer $\rightleftarrows$ .... $\rightleftarrows$ n-mer indefinite type of association as opposed to the monomer $\rightleftarrows$ n-mer micellar equilibrium usually observed for normal aqueous micelle systems. As a result, such reversed

TABLE IV. Structure and Characteristics of Some Surfactant Vesicle
Systems in Aqueous Solution[a]

Surfactant Structure	$10^{6}\overline{M}_{w}$[b]	R_{H}^{c}, Å	N^{d}	T_{c}^{e}
CATIONIC TYPE:				
$(CH_3)_2 R_1 R_2 N^+ X^-$				
if $R_1=R_2=C_{18}$, X=Cl, DODAC	13.4	400	48,500	36°C
if $R_1=R_2=C_{12}$, X=Br, DDDAB	7.0	---	---	--
ANIONIC TYPE:				
$[CH_3(CH_2)_{11}CO_2CH_2][CH_3(CH_2)_{11}CO_2]CHSO_3^-Na^+$	---	---	---	--
$(C_8H_{19})(C_7H_{15})C(H)(C_6H_4SO_3^-Na^+)$	23.0	260	5,760	--
NONIONIC TYPE:				
$[CH_3(CH_2)_{15}O]_2P(O)OH$, DHP	30.0	600	---	--
$[CH_3(CH_2)_{11}OCH_2]_2CHO(CH_2CH_2O)_{15}H$	12.0	---	---	none
ZWITTERIONIC TYPE:				
$[(CH_3)_3(CH_2)_{11}]_2N^+(CH_3)(CH_2)_3SO_3^-$	25.0	---	---	--
$[CH_3(CH_2)_{17}]_2N^+(CH_3)(CH_2)_2OP(O)_3^-$	17.0	---	---	38°C

[a]Data taken from References (15,31-35). [b]Refers to weight-average
molecular weight. [c]Refers to the hydrodynamic radius. [d]Number of
surfactant molecules per vesicle aggregate. [e]Phase transition temperature.

micellar systems do not exhibit a clear-cut CMC value as do normal micelles. Instead, at each surfactant concentration level, there is a distribution of aggregates and increases in the concentration lead to formation of larger aggregates in greater proportions (15,27). These Type I reversed aggregates are thus polydisperse, their average aggregation number is typically small (3≤N≤10), and they are postulated to have lamellar type structures in some instances. The aggregation number and size of such reversed micelles can be significantly altered by the amount of co-solubilized water present (41).

The Type II reversed micellar systems (i.e. those formed from anionic surfactants such as arylsulfonates or arylphenolates) exhibit aggregation behavior quite similar to that of normal aqueous surfactants. That is, they have fairly well-defined CMC values and much larger aggregation numbers compared to the Type I systems just described. Their aggregation number and size are, however, also dependent upon the water content and reach constant limiting values under specified experimental conditions (37-41). Due to the fact that these Type II aggregate systems are less complex (in terms of the number of actual species present) compared to the Type I systems, they have been touted as being the preferred system of choice in any separation science application (42). However, as will be shown from the applications in the literature, both Type I and II reversed micelles may be equally successfully employed (5).

The structure of some reversed micelle-forming surfactants as well as data on their aggregation behavior in different nonpolar solvents is presented in Table V. As can be seen, aggregates can form at very low surfactant concentrations in some cases and the size of the organized assemblies depends strongly on the amount of water present in most instances. Consequently, it is very important to stipulate both the surfactant and water concentrations when employing reversed micellar systems in separation science so that reproducible results are obtainable.

Lastly, mention should be made of surfactant microemulsions. Depending upon the relative concentrations, three component systems containing a surfactant, water, and a nonpolar solvent can form microemulsions (15,36,43). The addition of increasing amounts of an organic solvent (oil) to aqueous normal micellar solutions or increasing amounts of surfactant-entrapped water to reversed micellar solutions can lead to the formation of oil-in-water (o/w) or water-in-oil (w/o) microemulsions, respectively. Although potentially useful, there have been very few reports of their utilization in separation science (1,8). Such systems have, however, been successfully employed in a variety of industrial and related processes including enhanced oil recovery which is akin to a separation process (15,37,44,45). Due to space restrictions, the utilization of surfactant microemulsions in chemical separations will not be extensively discussed in this review article. The interested reader is referrred to many fine references on the properties and utilization of this type of organized surfactant system (15,36,43,46,66,67,215, 216).

TABLE V. Summary of Some Surfactants which Aggregate in Apolar
Solvents

Surfactant Structure (Abbreviation)	Bulk Solvent	Concentration Range^a	$\bar{N}^b$
Cationics of General Structure $R_1R_2R_3R_4N^+X^-$:^{c}			
$R_1=R_2=R_3=R_4=C_{14}$, X= $CH_3(CH_2)_2COO^-$; tetradecylammonium butyrate, TDAB	Benzene, with water added: g/5 mL: none 0.5 1.3	--- --- ---	4.3 23.2 34.5
$R_1=R_2=R_3=R_4=C_4$, X = ClO_4^- ; tetrabutylammonium perchlorate, t-BAP	Benzene	$10^{-3}-$ $10^{-2}m$	3-6
$R_1=R_2=R_3=C_{12}$, R_4=H, X= NO_3^- or HSO_4^- ; tridodecylammonium salts, TLAB or TLAN	Benzene	---	2-6
$R_1=R_2=R_3=C_8$, R_4=H, X = HSO_4^- or SO_4^{2-}; trioctylammonium salts, TOAB or TOAS	Benzene	0.4 - 8.0 wt%	1- 3.8
$R_1=R_2=C_{12}$, $R_3=R_4=CH_3$, X=Cl$^-$; didodecyldimethylammonium chloride, DDAC	Benzene	1.1 mmole/ kg	6.5 (at 50°)
$R_1=C_{16}$, $R_2=R_3=R_4=CH_3$, X=Cl$^-$; hexadecyltrimethylammonium chloride, CTAC	CHCl$_3$	$X_{CTAC}=$ 0.003-0.04	3.7- 7.0
$R_1=C_{12}$, $R_2=R_3=R_4$=H, X=$CH_3CH_2COO^-$; dodecylammonium propionate, DAP	Benzene Dichloromethane CCl$_4$	2.0 x 10^{-3}M 0.02-0.04 M 0.023 M	4-5 6.0 4.0
$R_1=C_4$, $R_2=R_3=R_4=CH_3CH_2COO^-$; butylammonium propionate, BAP	Benzene Dichloromethane CCl$_4$	0.05 M 0.11 M 0.025 M	4.0 5.0 3.0
Anionics:^{d}			
of General Formula $R-C-C=O(CH_2)CH(SO_3^-)C=O-O-R$ M^+:			
R= 2-ethylhexyl, M = Na; sodium bis- 2-ethylhexylsulfosuccinate	Benzene CCl$_4$	2.0 x 10^{-3}M^e 6.0 x 10^{-4} M	13- 23 17

Surfactant Structure (Abbreviation)	Bulk Solvent	Concentration Range[a]	$\bar{N}$[b]
	Cyclo-hexane	1 - 3 wt %	45-65
	Decane	6.5 wt %	25-31
R= decyl, M = Na; sodium didecyl-sulfosuccinate of General Formula:	Benzene	0.4 - 2.8 wt%	9-16

Surfactant Structure (Abbreviation)	Bulk Solvent	Concentration Range[a]	$\bar{N}$[b]
$R=C[CH_3][CH(CH_3)_2][CH_2CH(CH_3)_2]$, M = Na, sodium dinonylnaphthalene-sulfonate, NaDNNS	Benzene	0.05 - 0.2 wt%	10
$R=C[CH_3][CH(CH_3)_2][CH_2CH(CH_3)_5]$, M = Na, sodium didodecylnaphthalene-sulfonate, NaDDNNS	Benzene	0.5 - 2.8 wt %	9.7
	Decane	0.5 - 2.8 wt%	15.2
1,5-Dinonylnaphthalene-4-sulfonic acid[f], DNNSA	Benzene	2.0×10^{-5} M	3-12
	Hexane	2.0×10^{-5} M	7.0
	Toluene	2.0×10^{-5} M	6.0
Magnesium dilaurate, MgDL	Benzene	3.5 - 7.5 wt%	16.6
Lithium decanoate, LiD	Benzene	0.1 - 0.6 wt%	52 - 63
Nonionics:			
Sorbitan monooleate, Span-80, SP-80	Benzene	---	26
Polyoxyethylene(9.5)-t-octylphenol, Triton X-100, TX-100	CCl_4	0.32 M	1-4
Polyoxyethylene(6)nonylphenol, Igepal CO-530, I-CO530	Cyclo-hexane	0.04 M	----
Polyoxylene(20) sorbitan monolaurate, Tween 20, T-20	Benzene	---	3-15
	Chloroform	---	1-7
10 Oleyl Ether, Brij-96, B-96	Octane	---	----

Continued on next page

Table V. Continued

Surfactant Structure (Abbreviation)	Bulk Solvent	Concentration Range[a]	$\bar{N}$[b]

Zwitterionics:

of General Formula $RNH_3^+\ {}^-O_2CR'$ (where

$R \approx R'$ containing more than 8 carbon atoms):

$R=C_{12}$, $R'=C_{11}$; dodecylammonium

decanoate, DAD   Benzene 3.5×10^{-2}m   ----

of General Formula:

$$\begin{array}{l} CH_2O \cdot CO \cdot R \\ | \\ R' \cdot COO \cdot CH \quad\quad O \quad\quad\quad\quad CH_3 \\ | \quad\quad\quad\quad \uparrow \quad\quad\quad\quad / \\ CH_2O-P-O \cdot CH_2 \cdot CH_2 \cdot \overset{+}{N}-CH_3 \\ \quad\quad | \quad\quad\quad\quad\quad\quad\quad\quad \backslash \\ \quad\quad O^- \quad\quad\quad\quad\quad\quad\quad CH_3 \end{array}$$

Lecithins (phosphatidylcholines)	Benzene	0.001 - 0.01 wt %	80
	Benzene	0.7 - 1.0 wt%	73
	Chloro-form	---	68

[a]Refers to the operational CMC in most cases; i.e. concentration range where reverse micelles are present in the indicated solvent.
[b]Refers to the number average aggregation number in specified surfactant concentration range. Values were taken from references 5, 12-16, 27, 37-40.

[c]Typically exhibit Type I aggregation behavior. [d]Typically exhibit Type II aggregation behavior.
[e]Taken from reference 41. [f]Aggregation data is for the sodium salt under basic conditions.

Relevant Properties of Organized Surfactant Media. The addition of
surfactant to a solvent at surfactant concentrations/conditions under
which no aggregated species are present will usually not lead to any
appreciable alteration in the properties or processes occurring in the
solvent aside for possible salt effects upon the process and/or ion
pair formation between the surfactant and solute molecules. However,
the presence of organized surfactant assemblies can alter the
solubility of solutes, alter chemical and photophysical pathways and
rates, alter the effective microenvironment about solubilized solutes,
alter encounter probabilities in fast reactions, modify the position
of equilibrium processes, and alter the solution properties
(viscosity, surface tension, etc.) among other effects compared to
that of the bulk solvent in the absence of aggregates (1-16). Since
such organized surfactant systems mimic certain aspects of
biomembranes, they have also been referred to as membrane mimetic
agents (36,47).

     Although all of the mentioned properties of organized surfactant
media can potentially aid the separation scientist, in most cases, the
crucial factor in their successful application in separations is their
ability to selectively solubilize and interact with solute molecules.
The presence of surfactant micelles or vesicles can dramatically
enhance the solubility of a given solute compared to that in the bulk
solvent alone (1,4,16,40,41). For example, the presence of surfactant
inverted micelles allows one to solubilize polar species (salts,
bases, acids, water) in an organic solvent. Whereas the solubility of
water in alkane solvents like heptane, octane, or nonane is in the
range of 0.01 wt %, homogeneous mixtures of approximately 10% water in
these solvents can be prepared in the presence of reversed micelles
(such as in 0.015 M AOT) (48). Likewise, normal aqueous micellar
media can be employed to enhance the water solubility of organic
materials. For instance, 1,2-benzphenanthrene and 2,3-benzphen-
anthrene are virtually insoluble in water (water solubility $\leq$ 9.0 x
10^{-9} M). However, in the presence of 0.50 M potassium dodecanoate,
their solubility is roughly 6.4 x 10^{-4} M (16). This represents a
solubility enhancement of 66,000! Many other examples of such
enhancements in solubility are reported in the literature
(1,4,16,40,41).

     Depending upon the nature of the solute and organized surfactant
system, a solute can "bind" different regions of the aggregate system.
Figure 3 shows some of the solubilization sites available for a solute
in an aqueous normal micellar system (49). In inverted micellar
media, polar solutes can be solubilized in the interior water pool, or
associate with the headgroup of the surfactant molecule (if of
opposite charge). Additionally, less polar species can align
themselves with the surfactant molecules via both hydrophobic and
electrostatic interactions. The partitioning of a solubilizate (S)
between the bulk solvent (sol) and organized surfactant (sur) phase is
a dynamic equilibrium process with the degree of partitioning defined
by a partition (or distribution) coefficient P. The partition
coefficient is defined as the ratio of the solute concentration in

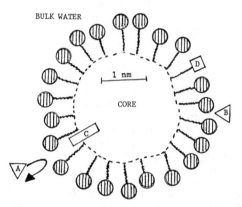

Figure 3. Simplified cross section of an aqueous normal micelle showing possible solubilization sites. A charged solute (A) would be electrostatically repelled from the micelle surface if it were of the same charge-type as the ionic micelle while an oppositely charged solute (B) would be electrostatically attracted to the micellar surface. Nonpolar solutes (C) would partition to the outer part of the more hydrophobic core region. Amphiphilic solutes (D) would attempt to align themselves so as to maximize the electrostatic and hydrophobic interactions possible between itself and the surfactant molecules. "Reproduced with permission from Ref. 49. Copyright 1984, Elsevier ."

$$P = \frac{[S]_{\text{in aggregated surfactant phase}}}{[S]_{\text{in bulk solvent phase}}} \qquad (1)$$

the organized surfactant assembly phase to that in the bulk solvent phase (equation 1) (1). In dilute solutions, the partition coefficient, P, can be related to a solute–surfactant aggregate binding constant, K_b, by use of the Berezin equation (50) (equation 2) in which $\bar{v}$ is the molar volume of the surfactant in the organized surfactant medium. The binding constant, K_b, for the interaction

$$K_b = (P - 1)\bar{v} \qquad (2)$$

of solubilizate with aggregate (equation 3) is merely the ratio of

$$S + sur \underset{}{\overset{K_b}{\rightleftharpoons}} S \cdot sur \qquad (3)$$

the concentration of the solute associated with the organized assembly [S·sur] divided by the free equilibrium concentrations of the uncomplexed solute [S] and surfactant aggregate [sur], respectively (equation 4). The [sur], sometimes designated C_m, is given by

$$K_b = \frac{[S \cdot sur]}{[S][sur]} \qquad (4)$$

the difference between the total surfactant concentration (C_T) and the critical concentration divided by the aggregation number of the surfactant assembly (1,5). Since the association rate of most solutes with surfactant aggregates is constant ($\approx 10^9 - 10^{10} M^{-1}s^{-1}$) (216), the larger the solute – aggregate binding constant, K_b, the more stable is the associated solute – surfactant aggregate complex and the longer is the solute's residence time in the organized surfactant assembly environment (termed pseudophase).

Tables VI and VII present some representative data on the binding constants and partition coefficients reported for the interaction of selected solutes with different surfactant micellar systems. The strength of the association of solutes with surfactant micelle assemblies is dictated by the net electrostatic, hydrogen-bonding, and/or hydrophobic interactions possible for a given solute – micelle combination under the prevailing experimental conditions. Consequently, as can be seen from the data in the Tables, the charge-type and chain length of both the solute and the micelle-forming surfactant as well as presence or absence of additives are important factors which can influence the magnitude of the binding constants (or partition coefficients). For instance, within a given family of solutes (such as the polycyclic aromatic hydrocarbons or quinones in Table VI or alcohols in Table VII), the degree of partitioning/binding to the micellar entity increases with increases in the solute hydrophobicity. Metal ions can electrostatically interact with and bind to anionic charge-type surfactant assemblies but not cationics (refer to entry for copper(II) in Table VI). For ionizable solutes, both hydrophobic and electrostatic interactions are

TABLE VI. Comparison of some Binding Constants for the Interaction
           of Solutes with Selected Organized Surfactant Systems

Solute	Organized Surfactant Assembly	K (M^{-1})	Ref.
2-Methyl-1,4-naphtho-quinone (menadione)	aq. NaLS micelles	1.2×10^4	51
2,3-Dimethyl-1,4-naph-thoquinone	aq. NaLS micelles	2.6×10^4	51
Duroquinone	aq. NaLS micelles	1.3×10^4	51
	BHAC reversed micelles in benzene[a]	$3.5 - 4.4$[b]	52
Naphthalene	aq. NaLS micelles	2.0×10^4	1,53
Anthracene	"	4.0×10^5	
Pyrene	"	1.7×10^6	
1-Methylquinolinium ion	aq. NaLS micelles	4.8×10^4	53
10-Methylacridinium ion	"	1.4×10^5	
Silver (I) ion	aq. NaLS micelles	1.3×10^3	53
Nickel (II) ion	"	2.4×10^3	
Copper (II) ion	aq. NaLS micelles	245[c]	54
	aq. DTAC micelles	0.002[c]	
Hydrogen ion (H^+)	aq. NaLS micelles	13.8	55
Copper-benzoyl-acetone complex	aq. NaLS micelles	4.9×10^3[c]	54
	aq. DTAC micelles	11.8[c]	
p,p'-DDT[d]	aq. CTAOH micelles[e]	1.5×10^3	56
	+ added BuOH[f]	1.8×10^3	
	+ added HexOH[f]	3.0×10^3	
	+ added KBr[g]	2.7×10^4	
	+ added KBr[g] & HexOH[f]	5.0×10^5	

[a]BHAC = hexadecylbenzyldimethylammonium chloride. [b]Refers to the
equilibrium constant (dm^3/mol) for distribution of solute between
the reversed micellar water pool and the bulk organic phase.
[c]Equilibrium constant (dm^3/mol) are given on a per monomer basis
(54).
[d]DDT = 1,1,1-trichloro-2,2-bis(p-chlorophenyl)ethane.
[e]CTAOH = Hexadecyltrimethylammonium hydroxide.
[f]Amount of alcohol added is $\leq$ 0.07 M. [g]Amount of KBr added is
$\leq$ 0.10 M.

TABLE VII.   Summary of Partition Coefficients for the Distribution
of Solutes between Normal Micellar and Aqueous
Pseudophases

Solute	Aqueous Normal Micellar System	Partition Coefficient	Ref.
1-Heptanol	NaDC	2000	207
	NaLS	2650	57
2-Heptanol	"	1500	
3-Heptanol	"	1010	
4-Heptanol	"	930	
1,8-Octanediol	NaLS	311	57
1,9-Nonanediol	"	743	
1,10-Decanediol	"	3800	
1-Pentanol	NaDC	100	207
	NaLS	820	58
	NaDeS[a]	650	
	SFONa[b]	535	
	NaLS/SFONa[c]	755	
	NaLS/SFONa[d]	1200	
Chloropentaammine cobalt(III)	NaLS	1.5×10^4	59
Propranol	CTAB	0.43	60
Penthianatemethobromide	CTAB	0.24	

[a]NaDeS = sodium decylsulfate.   [b]SFONa = sodium perfluorooctanoate.
[c]Mixed micelle in which the mol fraction of SFONa is 0.50.
[d]Mixed micelle in which the mol fraction of SFONa is 0.147.

possible. For example, the binding constants for the interaction of
protonated and unprotonated p-methylthiophenol with cationic CTAB
normal micelles are 1.0 x 10^3 and 8.3 x 10^3 M^{-1}, respectively (61).
The larger binding constant for the thiophenolate ion merely reflects
the additional electrostatic contribution to the binding interaction
compared to that possible for the neutral thiophenol. The addition of
additives (such as salt or alcohols) can also influence the magnitude
of the binding interaction (refer to data on DDT in Table VI).

To summarize, the binding interaction observed (or desired in
particular separation application) for a specific solute with a
surfactant assembly can be controlled by (1) variation of the
surfactant concentration (equations 3 and 4), (2) variation of the
charge-type and/or carbon chain length of the surfactant (refer to
data on duroquinone, Table VI and 1-pentanol, Table VII), and (3)
addition of appropriate additives (refer to DDT data in Table VI). By
manipulation of the experimental conditions just mentioned, it is
possible to observe differences in the binding/partitioning for
different families of solutes (Table VI) as well as for positional
isomers (see data in Table VII on heptanol isomers) with organized
surfactant media. In addition, differences in the binding of
enantiomers have been observed in a few cases (15,62). The fact that
one can utilize different sufactant organized assemblies to
differentially solubilize and bind a variety of solute molecules
serves as the main basis for their successful use in separation
science (1,5). Additionally, some of the other previsouly mentioned
unique properties of surfactant solutions and organized surfactant
systems can be judiciously exploited in order to aid the separation
scientist in some specific applications as will be detailed in later
sections of this overview.

Different Uses and Exploitation of Surfactant Systems in Separation Science

Organized surfactant assemblies have found amazingly diverse and
numerous practical applications in many areas of separation science.
Space limitations preclude an exhaustive review of all such systems
and applications. Consequently, only certain representative examples
will be given in many instances to illustrate the current
state-of-the-art, with emphasis given to the more recently developed
techniques. Potential areas for further research and future
developments will be identified. The main topics to be covered
include: use of surfactants as mobile phase additives and/or
stationary phase modification reagents in chromatographic separations,
with emphasis on micellar liquid chromatography; micellar
electrokinetic capillary chromatography; surfactant mediated
solubilization and extraction schemes; surfactant enhanced detection
schemes in separation science; a brief description of some
miscellaneous applications of surfactants; and lastly, a section on
some experimental considerations including surfactant and/or solute
recovery in surfactant-mediated separations.

Surfactant-Mediated Chromatographic Separations. The selective
interaction of surfactants with a variety of solutes (as ion pairs
with monomeric surfactant molecules or as bound ("associated") species

with micellar, vesicular, or liquid/crystalline organized surfactant
media) enables them to be applied in chromatography. From an
operational viewpoint, there can be two types of approaches to their
use in such chromatographic separations. First, they can be employed
in chromatographic mobile phases. An eluting solvent (mobile phase)
contains the surfactant(s) and solutes distribute between the
stationary phase (usually surfactant modified) and the surfactant (or
surfactant aggregate) in the mobile phase (Figure 4). Alternatively,
surfactants or surfactant organized assemblies can be immobilized in
(or onto) a stationary phase. Solutes are thus distributed between a
conventional mobile phase and the surfactant-modified stationary phase.

   Use of Surfactants in Chromatographic Mobile Phases. (1) Planar
and High-Performance Liquid Chromatography. Perhaps the most recent
development concerning the utilization of surfactants in
chromatography concerns their use as LC micellar mobile phases (1-8).
Surfactants had previously been successfully employed as mobile phase
additives in so-called ion-pair (68-73,118), soap (74), hydrophobic
(75,173), dynamic soap (76), ion interaction (77), hetaeric (78),
detergent-based cation-exchange (79) or surfactant chromatography
(69). There is still considerable debate concerning the retention
mechanism in this particular separation mode employing surfactants as
additives in the mobile phase (68,69,71,72). For these applications,
the surfactant concentrations and/or conditions are such that no
micellar aggregates form. That is, surfactant concentrations are
below the CMC value or conditions (high concentrations of added
alcohols) are such that micelles do not form. In fact, deviations
from the expected ion-pair retention behavior observed at higher
surfactant concentrations is usually attributed to micelle or mixed
micelle formation (71,79-82,118,121,173). Further information on the
use of surfactants as ion-pairing reagents in chromatography are given
in several fine reviews (69,81,83) as well as in a Chapter by Mullins
in this Symposium Volume (84).

   The first intentional use of surfactants in chromatographic
mobile phases at concentrations above the CMC was proposed in 1977 by
Armstrong and co-workers (1,86-100). Since the initial reports, the
general method, dubbed pseudophase liquid chromatography (PLC) or
micellar liquid chromatography (MLC), has moved from the realm of an
academic novelty to a demonstrated practical separation technique.
The basis for separation employing micellar mobile phases stems from
their ability to differentially solubilize and bind structurally
similar solutes. Skeptics view MLC as a fascinating example of the
incorporation of secondary equilibria for control or adjustment of
retention (101). However, it is the ultimate of secondary equilibria
since the types of interactions possible with micellar aggregates
cannot be duplicated by any single other equilibrium system, or for
that matter, any one or mixture of traditional normal or reversed
phase mobile phase systems. This is due to the fact that solutes can
interact with the surfactant aggregates via hydrophobic,
electrostatic, hydrogen bonding, and/or a combination of these
factors.

   A micellar mobile phase can be viewed as being composed of both
the surfactant micellar aggregates (pseudophase) and the rest of the

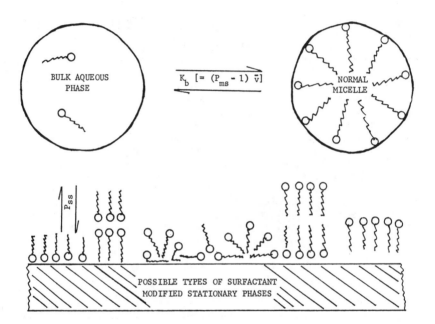

Figure 4. Artistic representation of the species and equilibria present when employing surfactant micellar mobile phases in LC.

bulk solvent (Figure 4). A solute thus distributes between the bulk
solvent – surfactant modified stationary phase (P_{ss}) and between the
bulk solvent – micellar pseudophase (P_{sm}). Consequently, there are
two partition coefficients which the separation scientist can try to
manipulate in order to achieve a desired separation. The basic
formulas relating these two partition coefficients and retention (in
terms of the reciprocal of the capacity factor) to the micelle
concentration are given in equations 5 and 6 for TLC and HPLC,
respectively:

$$\frac{R_f}{1 - R_f} = \frac{1}{\phi}\left[\frac{(K_b)\,C_m}{P_{ss}} + \frac{1}{P_{ss}}\right] \tag{5}$$

$$\frac{1}{k'} = \frac{1}{\phi}\left[\frac{(K_b)\,C_m}{P_{ss}} + \frac{1}{P_{ss}}\right] \tag{6}$$

where R_f and k' are the retardation and capacity factors,
respectively; ϕ is the phase ratio (equal to V_s/V_o where V_s and V_o are
the stationary-phase and void volumes, respectively); C_m is the
micelle concentration [equal to $(C_T-CMC)/N$ where C_T is the total
surfactant concentration]; K_b is the micelle-solute binding constant
(equal to $(P_{sm} - 1)\bar{v}$, where P_{sm} is the partition coefficient for
distribution of the solute between the micellar and bulk solvent
phases); and P_{ss} is the partition coefficient for distribution of the
solute between the bulk solvent and stationary phases (1,96,98,102).
These equations can be employed to describe or predict the retention
behavior exhibited by neutral solutes or ionizable solutes, provided
that there is only one form of the solute present over the surfactant
concentration range examined at a particular pH (in the latter case,
it would be necessary to bear in mind that the K_b term must be for the
actual form of the species present).

In the case of an ionizable solute where both the acid and
conjugate base (or base and conjugate acid) forms are present, the
following equation predicts the dependence of k' upon pH (at constant
surfactant concentration) or surfactant concentration (at constant
pH):

$$k' = \frac{k'_o[1 + K_b(C_m)] + k'_{cb}[1 + K_{bc}(C_m)]K_i/[H^+]}{1 + K_b(C_m) + [1 + K_{bc}(C_m)]K_i/[H^+]} \tag{7}$$

where K_b and K_{bc} are the binding constants for the interaction of the
weak acid and its conjugate base, respectively; k'_o and k'_{cb} are the
limiting capacity factors of the weak acid and its conjugate base,
respectively; C_m is the micelle concentration as previously defined;
and K_i is the apparent ionization constant for the weak acid (103). A
similar equation can be derived for weak bases and their conjugate
acids (103). It should be noted that equations 5, 6, and 7 can be
re-expressed in terms of several other chromatographic parameters or
by use of partition coefficients rather than binding constants (1).

The derivation of these different retention equations is important in several respects. First, they allow for calculation of micelle–solute binding constants, parameters which are important in many areas of micellar kinetics or chemistry. There have been several reports in the literature demonstrating this chromatographic approach for determination of micelle – solute binding constants (1,8,104,105). More importantly, they allow for prediction of retention behavior as a function of surfactant concentration (or of pH at constant micelle concentration), provided that the micelle – solute binding constant (or solute ionization constant) is known (which can be determined spectroscopically or from kinetic studies) (1,96,102). Consequently, the theory allows the chromatographer to determine the optimum conditions required for a desired separation.

Examination of equations 5, 6, and 7 reveals that retention can be controlled by variation of the surfactant micelle concentration, variation of pH (for ionizable species), and by manipulation of the solute-micelle binding constant (K_b) which, in turn can be influenced by additives (salt, alcohol; refer to data on DDT, Table VI) or the type (charge and hydrophobicity) of micelle-forming surfactant employed (refer to data in Table VII for 1-pentanol). Table VIII summarizes some of the factors that influence retention for surfactant-containing mobile phases and compares the effect of changes in these factors upon the retention behavior observed in both micellar liquid and ion-pair chromatography (81).

In addition to the factors listed in Table VIII, the nature of the surfactant-modified stationary phase affects P_{ss} (partition coefficient for distribution of solute between bulk solvent and modified stationary phases) and thus will influence the retention observed. It should be realized that most of the normal and reversed-phase packing materials will adsorb/absorb surfactant molecules from the mobile phase solution and become coated to different degrees when surfactant mobile phases are passed through them. Numerous adsorption isotherms have been reported for various surfactant – stationary phase combinations illustrating this point (82,85,106,115-128,206). The presence of additives can mediate the amount of surfactant surface coverage obtained (110-129,175,206). It has been postulated that the architecture which adsorbed surfactant molecules can assume on conventional stationary phases can range from micellar, hemi-micellar, or admicellar to mono-,bi-, or multilayered, and/or other liquid crystalline-type structures (93,106,124,128,129, 132,208,212,217,221). In a few cases, it has been reported that there can be a relatively slow reorganization of the stationary phase surfactant structure (137) and similar ageing (storage) effects on the micellar aggregate structure in solution have been noted as well (138). Also, the structural parameters (i.e. pore diameter and partical size distribution) of the stationary phase packing material can be altered due to surfactant adsorption (133-135). The fact that many stationary phase properties are substantially altered by the process of surfactant adsorption has important implications with regard to chromatographic retention and efficiency (93,106,110,126-128). A review on the role of the stationary phase in MLC is given in a Chapter by Berthod, et al in this Volume (136).

TABLE VIII. Comparison of the General Effect of Variables on
Retention in Reversed-Phase Ion-Pair (RP-IPC) and
Micellar Liquid Chromatography (MLC)

Factor Varied	Effect upon Retention	
	in RP-IPC[a]	in RP-MLC
Concentration of surfactant of mobile phase	Increasing concentration increases retention (up to a limit)	Increasing concentration decreases retention (down to a limiting value)[b]
Presence of an organic modifier (added alcohol or acetonitrile)	Retention decreases with increasing concentration or hydrophobicity of the organic additive	Same as in RP-IPC[c]
pH	Retention increases as pH manipulation maximizes the concentration of the ionic form of the solute	Depends upon the nature (i.e. charge-type and concentration) of the surfactant micelle and ionizable solute; eq. 7 predicts a sigmoidal-type dependence between retention and pH (at constant surfactant concentration)[d]
Temperature	Retention increases as temperature decreases	Retention decreases slightly as temperature increases[e]
Ionic Strength		Rention decreases as ionic strength increases[f]; linear dependence between k' vs. $\mu^{-1/2}$

[a]Information taken from Ref. 81. [b]See references 1,96,102,105,106, 114,121,131.

[c]See references 106-112,121,130,131,154,201; some exceptions with very short-chain alcohols (MeOH) (110). Table IX shows the effect of alteration of alcohol hydrophobicity in MLC retention (112).

[d]See reference 103. [e]Refer to reference 113, Figure 2.

[f]Refer to references 98,99,107,108,110,130,131). There are some apparent notable exceptions to the general trend, see, for instance, Ref. 98,99,103,110.

TABLE IX. Effect of Added Alcohols upon the Chromatographic
           Retention and Efficiency of 2-Ethylanthraquinone using
           a Micellar Sodium Dodecylsulfate Mobile Phase and a C-18
           Reversed Phase Column[a]

Additive	Capacity Factor k'	N
none[b]	37.1	50
5% added methanol	27.3	56
5% added ethanol	20.6	100
5% added n-propanol	13.2	320
5% added n-butanol	9.5	725
2% added n-pentanol	12.3	---
5% " "	7.3	810
5% added DMSO	22.1	---

[a]Temperature 23.5° C, data taken from Ref. 112.

[b]The micellar mobile phase in all experiments consisted of aqueous
0.285M NaLS, flow rate 1.00mL/min, 10-cm column.

From an experimental standpoint, it is important to properly equilibrate the column with the surfactant mobile phase prior to use so that reproducible chromatographic results can be obtained.

In terms of chromatographic applications, the advantages of employing surfactant micellar mobile phases that have been cited include: enhanced selectivity, low cost, low toxicity, ease of mobile phase disposal, ease of purification of the mobile phase (i.e. water and surfactant), and the ability to simultaneously chromatograph both hydrophilic and hydrophobic solutes among others (1-3,88,95,97,111). More recently, several other unique chromatographic advantages obtainable have been reported (139). First, use of some micellar mobile phases allows for more convenient and rapid gradient elution (i.e. gradient in terms of micellar concentration) compared to that possible with conventional hydro-organic mobile phases (140,141). Secondly, it has been reported that utilization of reverse micellar mobile phases (AOT in hexane) in normal phase chromatography can greatly reduce or eliminate the solute retention dependence upon water content that is usually observed in normal phase LC (142). Third, the use of some micellar mobile phases allows for new or enhanced modes of detection in TLC or HPLC. More details on this point will be presented in a latter section of this review article. Some of these unique chromatographic capabilities of micellar mobile phases are discussed in more detail in a Chapter by Dorsey in this Symposium Volume (143).

A fourth major reason for employing such micellar phases in HPLC is that they allow for the direct injection of untreated biological fluids (urine, plasma, saliva) (144-149,218) as well as waste water samples (112). Thus, this technique is very useful in therapeutic drug monitoring since the micellar solution can solubilize the serum/urine preventing protein precipitation and displace the drug/analyte from the serum/urine components thus allowing the analyte to partition to the surfactant modified stationary phase (144-149). Consequently, minimal sample preparation is required and the analysis time is reduced. There will no doubt be further break-throughs in this fast-moving field in the near future with respect to novel chromatographic advantages of surfactant-containing mobile phases.

Lastly, the use of micellar mobile phases allows a convenient means of studying micelle - solute interactions (i.e. determination of binding constants) (1,104,105) as well as determination of surfactant CMC values (from breaks in the log k'$_{solute}$ vs. log C_T plots) (64,109,148,172). In this area, the more important application is its use in the determination of binding constants (1).

The main disadvantages of micellar chromatography are the observed diminished chromatographic efficiency, higher column back pressure, and in preparative work, the need to separate the final resolved analyte from the surfactant (95) (a later section of this review will discuss this latter problem and its resolution in further detail). The higher column back pressure and part of the decreased efficiency stem from the fact that surfactant-containing mobile phases are more viscous compared to the usual hydro-organic mobile phases employed in conventional RP-HPLC (refer to viscosity data in Table X)

TABLE X.   Viscosity of Commonly Employed Solutions in Micellar
           Liquid Chromatography[a]

Surfactant System	Viscosity, cP
Methanol alone	0.55
Distilled Water alone	1.01
0.10 M NaLS	1.21
0.10 M CTAC	1.31
0.10 M SB-12	1.16
0.10 M NaDC	1.32
0.27 M CTAB + 50% n-BuOH[b]	4.48
0.40 M NaLS	2.27
0.40 M CTAC	2.46
0.40 M SB-12	1.76
0.43 M CPC	3.40
0.04 M DODAB (Vesicle System)	5.2

[a]Data taken at 25.6° C; Reference 112.

[b]Probably a microemulsion system.

(106,112,118). Due to the relatively high viscosity of surfactant
vesicle and microemulsion systems (refer to data on DODAB and CTAB/50%
BuOH in Table X), their use in HPLC will be limited since lower flow
rates would be required which would lengthen the required time for a
separation. Additionally, most surfactant vesicular (112) as well as
some micellar solutions are optically opaque which limits the
wavelength range available for spectroscopic detection unless a
postcolumn dilution step is employed (219).

The major contributions which result in the reduced
chromatographic efficiency have been ascribed to slow mass transfer
principally due to poor wetting of the surfactant modified stationary
phase (109), poor mass transfer between the micelle and stationary
phase (113), and poor mass transfer in the stationary phase (100,106).
In some cases, the use of small amounts of alcohol additives (MeOH,
n-PrOH) and operation at elevated temperature (40° C) result in
chromatographic efficiencies comparable to that seen in traditional LC
using hydro-organic mobile phases (109,113,154,206). In our own work,
we have found n-pentanol to be superior to n-propanol in this regard
(refer to Table IX) (112). Further work is clearly needed in this
efficiency area in order to clarify the exact reason(s) for the
reduction in efficiency. It appears that a combination of factors can
contribute to this effect with the dominant efficiency reduction mode
dependent upon the nature of the solute, micellar mobile phase, and
stationary phase packing material employed (100,112,135).
Consequently, all explanations given to date are probably correct for
the particular limited cases examined in the work cited.

Micellar mobile phases have been utilized in numerous recent
paper, thin-layer, and high-performance liquid chromatographic
separations. Table XI summarizes the separations performed to date.
As can be seen, the general approach is amenable to separation of a
wide variety of organic, biological and inorganic species. It appears
to hold particular promise in the areas of metal/anion speciation
(156,162,163) and in biological/protein separations (137,165,170,
223). More details concerning application of micellar mobile phases
in the separation of organic and inorganic ions is presented in a
Chapter by Mullins in this Volume (84). A recent review by Matson and
Goheen (165) outline some of the considerations and applications of
utilizing detergent-micelle mobile phases in the HPLC separation of
membrane proteins. In many instances, the combination of several
chromatographic steps, one or more of which employed surfactant/
micelle mobile phases, has proven to be useful in the separation of
biological materials (166,167,171,223,224,233).

(2) Gel Filtration. Micellar solutions have also been utilized
in gel permeation (filtration) chromatography (1). In fact, the first
example of a separation which used a micellar mobile phase was in this
area of exclusion liquid chromatography (ELC) (86). The last six
entries in Table XI summarize some of the separations/work reported
concerning micellar mobile phases in ELC. In most of these
applications, the work was conducted with stationary phases of
relatively small pore size. With these type phases, the relatively
large micellar aggregates are confined to the excluded volume of the
column and elute rapidly whereas smaller solute molecules in a mixture

TABLE XI. Summary of Some Selected Separations Reported which have
          Utilized Surfactant-Containing Mobile Phases[a]

Component(s) Separated	Stationary Phase	Mobile Phase Composition	Mode	Ref.
Phenols (28)	Whatman No. 3 paper strips	Aq. NaLS or CTAB/ 8% PrOH	PC[b]	150
Dyestuffs (13), anions (2), cations (3)	Whatman No. 1 paper	CTAB or NaLS in 50%HOH/50% BuOH[c,d]	PC	164
Polycyclic aromatic hydrocarbons, pesticides (4)	polyamide or alumina sheets	Aq. NaLS or CTAB	TLC	87, 91
Nucleosides (4)	silanized silica gel-60	Reversed micelles of sodium dioctylsulfo-succinate in cyclo-hexane	TLC	87
Nucleosides (4)	polyamide	Aq. NaLS	TLC	87
Dyes, Pesticides	polyamide or alumina	Aq. NaLS or CTAB	TLC	151
Dyes, Food Colors	alumina or polyamide	Aq. NaLS	TLC	94
Mycotoxins	polyamide, alumina, or RP sheets	Aq. NaLS	TLC	1,168
Amino acids (3)	column	Brij-35/30% EtOH[c]	column	213
Proteins (6)	supelcosil LC-8	Aq. Neodel 91-6	HPLC	152
Hydroxybenzenes (phenols, quinols, catechols) (18)	C-18 RP	Aq. NaLS	HPLC	105
Dithiocarbamates (5)	Bondapak CN	Aq. CTAB/30% MeOH	HPLC	108, 206
Anions (5)	Spherisorb ODS	Aq. CTAC	HPLC	107
Test Mix	Polygosil ODS	Aq. NaLS	HPLC	153
Anthracyclines	Hypersil ODS	Aq. Brij-35[c]	HPLC	147
Nucleosides, bases	Polyvinyl-alcohol	Aq. 0.01M NaLS, pH 3.4	HPLC	218
Alkyl-benzenes, PAH's, phthalates, chlorinated benzenes	C-8 silica	Aq. 0.2M NaLS	HPLC	220

Component(s) Separated	Stationary Phase	Mobile Phase Composition	Mode	Ref.
3-Alkylbenzene-sulfonates	QAE-2SW (anion-exchanger)	Aq. DTAB	HPLC	222
Substituted benzenes (9), ethyl esters (5)	RP-18	Aq. SB-12	HPLC	128
Berberine-type alkaloids (4)	Bondapak-Ph	Aq. NaLS/30% MeOH	HPLC	154
Therapeutic drugs (9)	Supelcosil LC-18 or LC-CN	Aq. NaLS	HPLC	144
Tyrosinyl peptides (5), aromatic ketones (7)	Hypersil	60:40 Water:MeOH containing Tween-20 and NaLS; pH 3.08[c]	HPLC	76
Phenols (8)	Ultrasphere octyl	Aq. NaLS (gradient) pH 2.5	HPLC	155
Triglycerides (10)	Various RP C-18 MCH10, both end-capped and non	Aq. NaLS, Aq. CTAB	HPLC	160
Catecholamines (5), 1-phenyl-alkaylamines (4)	SP-2SW cation-exchanger	Aq. NaLS, pH 3.5 (or 4.6)	HPLC	169, 222
cis/trans Co(III) complexes[e]	methyl or phenyl bonded phases	Aq. CTAB	HPLC	156
Cu(II)/Ni(II)[f]	methyl or phenyl	Aq. NaLS/5% MeOH	HPLC	156
Zn(II), Pd(II), Cu(II)[g]	Radial pak-silica	Aq. NaLS	HPLC	163
Aromatic amino-sulfonic acids, nucleotides (9)	octadecyl-Spherisorb	Aq. SB-10/20% aceto-nitrile, pH 4.7 or 3.0	HPLC	130, 131
Phenols, PAN's	Micro-Pak MCH-10	Aq. NaLS	HPLC	95
t-RNA's (6)	Microparticulate bonded phases	n-Decylbetaine, pH 5.5 - 6.5	HPLC	130
Polypeptides (9), protonated phenylalanine oligomers (5)	C-18 end-capped	Brij-35 or Triton X-100/15-30% added acetonitrile[c]	HPLC	120
Bromhexine	Bondapak C-18	NaLS/25% MeOH[c]	HPLC	210
Drugs (5)	Supelcocil CN	Aq. Brij-35	HPLC	148

Continued on next page

Table XI. Continued

Component(s) Separated	Stationary Phase	Mobile Phase Composition	Mode	Ref.
Test mix (6), Vanillin/ethyl Vanillin	Radial-PAK C-18	Aq. Brij-35	HPLC	106
Determination of folylpolyglutamate hydrolase activity	PXS 10/25 ODS	Aq. 0.20M NaLS	HPLC	224
Thiols (8), nitrosoamines (9), and quinones (20)	C-18 or C-8 RP	Aq. NaLS, CTAB, CTAC normal micelles or AOT/cyclohexane reversed micelles	HPLC	112
Proline/hydroxyproline	Ultrasphere ODS	Aq. NaLS, pH 2.8	HPLC	162
t-RNA's (5)	Sephadex G-100-120	Aq. CTAB/NaCl, pH 8	GF[h]	86
Nucleosides, nucleotides (8)	Sephadex G-25	Aq. Sodium dodecanoate pH 8	GF	157
Amino acids (14)	Sephadex G-25	Aq. Sodium dodecanoate	GF	158
Nucleotides (5)	Sephadex G-25-300 or G-100-120	Aq. CTAB, pH 8.0	GF	159
Alkylbenzenes (4)	Hypersil silica	Aq. NaLS/2% BuOH	GF	193
Amino acids	Sephadex G-25	Aq. NaLS	GF	194

[a]Micellar mobile phases unless otherwise specified. [b]PC = paper chromatography. [c]Presence of micelles is unclear. [d]Most likely a microemsulion system. [e]As iminodiacetate complexes. [f]As N,N'-Ethylene-bis(acetylacetoneimine) chelates. [g]Separated as tetrakis(1-methylpyridinium-4-yl)porphine metal complexes. [h]GF = gel filtration.

can reside in the pore volume, thus requiring longer elution times.
However, if these smaller solute molecules can partition to the
micellar pseudophase and bind the micelle entity, then they will elute
more rapidly (1). Consequently, solutes can be separated based on
their differential binding ability to a particular micellar assembly.
Equation 8 shows the dependence of the elution volume, V_e,
corresponding to the maximum concentration in an emerging band, upon
the surfactant micelle concentration in the mobile phase equilibrating
the column (183):

$$\frac{1}{V_e - V_o} = \alpha K_b C_m + \alpha \qquad (8)$$

where α is an experimental constant (see references 1,159,182,183), V_o
is the excluded volume, and K_b is the micelle-solute binding constant
as previously defined (1,86,157,158,159,183). Via use of equation 8
or alternative re-expressed versions (1), the binding constant (or
partition coefficient) of different solutes to micellar systems have
been determined (159,182,183,226).

A cursory review of the literature reveals that the ELC technique
with micellar mobile phases has proven to be very beneficial in the
characterization of micellar systems (184-186,190-192,227,228). For
example, microcolumn exclusion LC has been applied to the
determination of the CMC value of surfactants (or micellar-forming
proteins), determination of the kinetic rate and equilibrium
association constants for surfactant (or protein) micellization
(184,192), determination of the size or size distribution of micelles
(especially those formed from block copolymers or milk casein)
(185,186,191,192,225) as well as for estimation of the time required
for formation of micelles (or micelle-forming macromolecules) (186)
among others. The size and stability of reversed micelles has also
been evaluated using ELC (195).

The use of ELC to characterize micellar and related aggregates
thus appears to be popular and useful. In fact, its use in this
manner overshadows the analytical applications of micellar mobile
phases to aid ELC separations. However, several recent reports do
point out the advantages of micellar mobile phases in ELC (187-189)
for the isolation and purification of bacterial and viral proteins.
For instance, bacteriorhodopsin solubilized in octylglucosides (OG)
was isolated at analytical and preparative levels from the denatured
protein and free retinal (187) and an influenza viral protein was
isolated using NaLS or Brij-35 eluents with TSK G3000SW or TSK G5000PW
columns (188). Other such applications will no doubt be forthcoming
in the near future. It has been reported that for micellar-mediated
ELC to develop into a viable technique requires "the development of a
high-performance GPC packing material that has an exclusion limit of
roughly 1,000 - 2,000 and is compatible with the aqueous micellar
mobile phase" (1). Future work should be directed in this area.

Surfactants as Stationary Phases. (1) Applications of
Surfactants "Immobilized" as a Stationary Phase. Apart from their use
as mobile phase additives, there are instances where surfactants have
been immobilized or coated on stationary phases, especially for

packing materials in GC, GLC, and column, paper or TLC (176-180,229-
231). Surfactants such as Aliquat 336, cetrimide, CTAB, NaLS, TX-100,
Surfynol 485, trioctylamine, etc. have been coated or deposited on
capillary, macroreticular resins, Whatman No. 1 paper, chromosorb P,
alumina, or silica supports. Borosilicate glass capillary columns in
which CTAB is electrostatically incorporated to the inner surface thus
forming a thin film of hydrophobic stationary phase for use in
capillary liquid chromatography have also been described (232). When
used in conjunction with surfactant CTAB mobile phases, efficient
separations of drugs from their metabolites are possible using such
open tubular columns (232,233).

The problem with using surfactant-modified stationary phases in
LC is that the surfactant will usually slowly elute (bleed) from the
support thus resulting in different retention behavior of solutes with
time. This is why most applications are in the area of GC or GLC. An
exciting recent advance has been reported by Okahata, et al (181).
Namely, a procedure has been developed for immobilizing a stable
surfactant vesicle bilayer as the stationary phase in GC. A bilayer
polyion complex composed of DODAB vesicles and sodium poly(styrene
sulfonate) was deposited on Uniport HP and its properties as a GC
stationary phase evaluated. Unlike previous lipid bilayers which
exhibited poor physical stability, the DODAB polyion phase was stable.
Additionally, the temperature-retention behavior of test solutes
exhibited a phase transition inflection point. The work demonstrates
that immobilized surfactant vesicle bilayer stationary phases can be
employed in GC separations (181). Further work in this direction will
likely lead to many such unique gas chromatographic supports and novel
separations.

(2) Micelles as a Liquid "Pseudo-Stationary" Phase -- Micellar
Electrokinetic Capillary Chromatography (MECC). MECC (also called
micellar capillary electroosmotic chromatography (205)) is a
separation technique first described by Terabe et al (196) which
combines many of the operational principals and advantages of micellar
liquid chromatography and capillary zone electrophoresis (196-205).
Solutes in a mixture (both ionic and neutral) are separated based on
their differential partitioning between an electroosmotically-pumped
aqueous mobile phase and the ionic surfactant micellar aggregate which
possesses an overall fractional charge and moves at a velocity
different than that of the aqueous mobile phase due to electrophoretic
effects. Thus, the separation mechanism is akin to that of
conventional liquid-liquid partition chromatography, with the micellar
entity functioning as a "pseudo-stationary" phase (197,201). Some
view MECC as an example of a laminar microscopic counter-current
separation technique (1). The fundamental characteristics and factors
effecting retention and efficiency in MECC have been described
(196,197,199,202,204,214). The approach results in excellent
resolution due to the very high efficiency obtainable (200,000 -
600,000 theoretical plates, HETP ca. 1.9 - 3.7 μm) (196,203). MECC
has been employed to separate a variety of environmental and
biological-type mixtures (see Table XII); including the analysis of
vitamins in spiked human urine (201). Electrokinetic measurements can
also be employed to evaluate surfactant critical micelle
concentrations (236).

TABLE XII.   Summary of Successful Applications employing Micellar
Electrokinetic Capillary Chromatography (MECC) in
Separations

Class of Compounds Separated	Micellar Solutions Employed/Conditions	Reference
Phenols (8), xylenols (6)	Aq. 0.05 M NaLS, pH 7.0	196
Amino acids (22) [as their phenylthiohydantoin derivatives]	Aq. 0.05 M NaLS or 0.05 M DTAB, pH 7.0	198
Chlorinated phenols (7)	Aq. 0.10 M NaLS, pH 7.0	199
Isomeric chloro-phenols (222)	Aq. 0.07 M NaLS, pH 7.0	199
Aromatic sulfides (11)	Aq. 0.02 or 0.05 M NaLS; 80:20(%) 0.03 M NaLS: MeOH pH 7.0	200
Metabolites of Vitamin B_6 (6)	Aq. 0.05 M NaLS, 0.01 M phosphate, 0.001 M borate	201
Substituted purines (6)	Aq. 0.05 M NaLS, 0.001 M borate, 0.01 M phosphate	202
Nitroaromatic compounds (4)	Aq. 0.01 M NaLS, 0.01 M phosphate	203
Metal ions [Mn(II), Co(II), Zn(II), Cu(II)] as their tetrakis(4-carboxyphenyl)-porphinato chelates	Aq. 0.02 M NaLS, 0.05 M phosphate, 0.0125 M borate	205
Oligonucleotides (7)	Aq. 0.05 M NaLS/3mM Mg(II)	235
Polythymidines (7)	Aq. 0.05 M NaLS/0.3mM Cu(II)	235

MECC is the most recent and fastest developing surfactant-mediated technique. Future work is required to extend the range of retention possible (197). Presently, the total elution range is relatively narrow. It should be possible to manipulate retention and improve separation at the two extremes of the elution range by judicious addition of additives (organics/salts) to the micellar surfactant solution (200,234,235). As previously mentioned, the presence of such additives can alter the partitioning (P_{sm}) of a solute between the aqueous and micellar phases. Alternatively, the use of other types of capillary materials (as alternatives to fused silica) or coatings of polymeric materials on the inner wall of the fused silica (214) may prove beneficial in this regard. More details on the current status of this separation technique are given in a review by Armstong (1) and a report by Sepaniak et al (203) in this Symposium Volume.

Surfactant-Mediated Solvent Extractions. Partitioning and extraction separation techniques serve to provide sample purification as well as a simple and effective means for improvement of analytical methods by enhancement of both sensitivity (by sample concentration) and selectivity (by removal of potential interferences). Although not well appreciated, many surfactant and micellar-mediated extraction systems have been described in the literature, especially in the area of metals analysis and in biological purifications (1,5,237,238). There are several different types of surfactant-mediated extraction schemes possible depending upon the nature of the analyte mixture and the extracting surfactant system employed. These can be broadly divided into two types: (1) those involving nonpolar solvent - surfactant (or reversed micellar) systems and (2) those involving aqueous surfactant/micellar media. In many instances, the possibility of w/o or o/w microemulsion formation in these systems also exists (239). Surfactants in organic solvents have been utilized to extract ions, complexes, and enzymes from aqueous or solid matrices. Likewise, some aqueous surfactant/normal micellar systems have been employed to extract biological, organic, or agricultural materials from other aqueous, organic, or solid matrices. Additionally, use of certain aqueous micellar media allows for concentration and purification of metal ions, organic compounds, or biological substances due to their phase separation behavior (i.e. cloud point phenomena or coacervation behavior). While there have been many practical applications using these surfactant systems in extractions, mechanistic studies have lagged behind due to the complicated nature of the physicochemical processes involved and lack of knowledge of the surfactant structures present under the extraction conditions. Since the rational design of future separation systems requires an understanding of the processes involved in these surfactant extraction procedures, future work should concentrate on the mechanism of such separations. Next, a brief description and illustrative examples of each of these types of surfactant-mediated extraction techniques will be given.

Extractions Utilizing Surfactants in Organic Solvents. The use of organic solvents containing surfactants in extractive metallurgy has probably been the most prevalent application of surfactants in chemical separations (1,5,240-262). Table XIII summarizes some of the

TABLE XIII.   Name and Structure of Several Different Types of
Extractants Utilized for Metal Ion Separations[a]

Name	Structure

Anion Exchanger Type:

Primene $\quad$ $(CH_3)_3C(CH_2C(CH_3)_2)_4NH_2$

Aliquat 336 $\quad$ $R_3N(CH_3)^+\ Cl^-$ where $R = C_8 - C_{10}$

Adogen 381 $\quad$ R_3N where R = isooctyl

Alamine 336 $\quad$ R_3N where $R = C_8 - C_{10}$

TOA $\quad$ R_3N where R = octyl

Adogen 283 $\quad$ R_2NH where $R = C_{13}$

Acidic Extractants:

Di-2-ethylhexylphosphoric acid $\quad$ $(C_4H_9CH(C_2H_5)CH_2O)_2PO_2H$

Versatic 10 $\quad$ R_3CCO_2H where $R = C_8$

Fatty Acids $\quad$ RCO_2H where $R = C_{14} - C_{18}$

SYNEX 1051

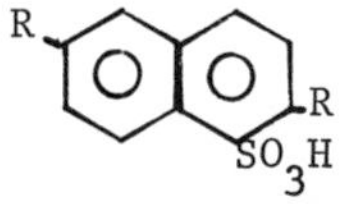

where $R = C_9H_{19}$

Solvating Extractants:

Tri-n-butylphosphate $\quad$ R_3PO where $R = C_4H_9O$

Trioctylphosphine oxide $\quad$ R_3PO where $R = C_8H_{17}$

Dihexylsulfide $\quad$ RSR where $R = C_6H_{13}$

Chelating Type Extractants:

Kelex 100

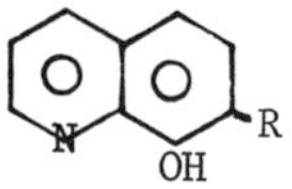

where R = dodecenyl

LIX 63 $\quad$ $C_4H_9CH(C_2H_5)CH(OH)C(NOH)CH(C_2H_5)C_4H_9$

LIX 34

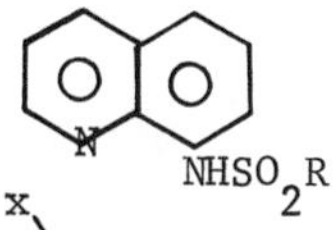

where R = p-dodecyl-
benzene

Polyols

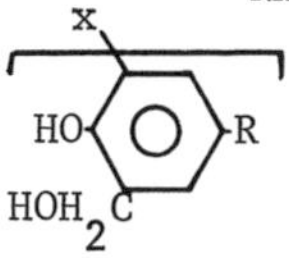

Continued on next page

Table XIII. Continued

Name	Structure
LIX 65N	where R_1 = phenyl, R_2 = H, and R_3 = C_9H_{19}
LIX 54	where R_1 = CH_3 and R_2 = p- or m-dodecyl

[a]Data taken from reference (264).

different types of extractants which have been utilized to extract
metal ions from aqueous solution into an organic layer containing
the extractant. As can be seen, many of these extractants are
surfactants. Depending upon the specific conditions and type of
analytes present, such surfactants can function as either ion-pair
or phase transfer agents (264-266), or exist in aggregated form as
reversed micelles (240,241,264) or in some cases, microemulsions
(239,251). Although still subject to debate, recent accumulated
evidence strongly supports the argument that reversed micelles are
present in the organic phase and play a vital role in many metal ion
extractions involving surfactant extractants such as those depicted
in Table XIII (240-266). For example, it has been recently shown
that reversed micelles of di-n-butylphosphate, quaternary alkylammonium salts, metal alkylarylsulfonates, alkylsulfates, dialkyl-
dithiophosphates, di(2-ethylhexyl)phosphoric acid, phenols, di-
nonylnaphthalene sulfonic acid, and SP-3 carbozoline can form in the
organic layer during metal ion extractive conditions (267-283).
Under some conditions, microemulsions form (284,285). In addition
to these surface active extractants, many extraction schemes also
have some other surfactants present (such as those given in Table V)
that can form reversed micelles in the organic phase (1,4,5,283,330).
The dialkylnaphthalene sulfonates (see Table V, anionic surfactant
section) have been especially useful in this regard (263).

Some of the organic-containing surfactant systems that have
been utilized in the extraction of a variety of metal ions (as
cations or metal complexes) from aqueous solution are summarized in
Table XIV (286-321). Inspection of the Table indicates that most of
the successful extraction schemes involve use of either cationic or
anionic surfactants/extractants. In contrast, only a relatively
few recent applications employed zwitterionic or nonionic surface
active agents. The interested reader is referred to several recent
monographs/review articles for more extensive compilations of different extraction systems involving surface active agents (330-332,
346,356,358). It should be emphasized that organic species that are
capable of being ionized can also be extracted from aqueous media or
solid matrices (refer to the last 4 entries of Table XIV) (322-325).
Lastly, mention should be made of the fact that many of the systems
given in Table XIV can not only be conducted at analytical or
preparative scales but also at the process level (356). Some of the
practical applications include recovery of metals from spent electro-
lytic or scrap leaching liquors (353), extractions from synthetic
mixed fission product solutions (354), and separation of rare earth
metals from ores (355). The general approach should also be
potentially useful in pyrometallurgical operations involving melts
or molten slags (if higher boiling organic solvents are employed).

In many of the examples presented in Table XIV, the existence
of reversed micelles (275,278,279,282,293,315,326,347) or microemulsions (349-352) is implicated and their presence is an important
factor which influences the characteristics of a particular extraction process. Often, quantitative descriptions of such extractions
is difficult due to the fact that many of the reversed micellar
systems formed undergo an indefinite type of self-association in

TABLE XIV.   Compilation of Selected Extraction Systems that Involve
             Use of Organic Solvents in Presence of Surfactants[a]

Component/Aqueous Conditions (Ref.)	Surfactant/ Organic Solvent	Additives/ Co- extractant	Comments
Nd, Tb, Tm (286, 287)	Di(2-ethylhexyl) phosphoric acid [HDEHP] in cyclo- hexane	none glycine 3-mercapto- propionic acid	α= 2.53[b] α= 2.85 α= 2.35
Cm,Cf (in presence of many other salts) (288)	HDEHP	none	α= 1.2 - 6.0
Ta,Nb in aqueous oxalic acid or HCl (289)	HDEHP in heptane	none	Eff= 85%[c]
Co, Ni (290)	HDEHP in xylene, n-dodecanol, or dodecane	none	-----
Lanthanides/ Actinides (291)	HDEHP in aromatic solvents	Dinonyl- naphthalene- sulfonic acid	-----
Zn in aqueous zinc sulfate (292)	HDEHP in kerosine	none	-----
Al, Ga, In (293)	Decanoic acid in benzene or octanol	NaClO$_4$	-----
Cu in aqueous NaClO$_4$ (294)	Decanoic acid in benzene or octanol	-----	-----
Th from other metals in acetic acid (295)	Versatic-10 in butanol	none	-----
Cu,Cd,Co,Ni from wastewater (296)	Palmitic, stearic, or linoleic acid in kerosine	none	-----
Zn,Co from aqueous solutions (297)	(t-dodecylthio) acetic acid in kerosine	none	-----
Fe from aqueous HNO$_3$ (298)	Tributylphosphate in kerosine	none	-----
Mo from aqueous HCl (299)	1,5-Bis(dioctyl- phosphinyl)pentane in chloroform	none	Eff= 93%
Pd in nitric acid (300)	Diheptylsulfide in benzene or chloro- form	none	-----

Component/Aqueous Conditions (Ref.)	Surfactant/ Organic Solvent	Additives/ Co-extractant	Comments
Lanthanides in aqueous NaClO₄ (301)	Didodecylnaphthalene sulfonic acid in toluene	none 4-t-butyl-cyclohexyl-15-crown-5	----- 20-50 % en-hance-ment[d]
Cu from other metals (302)	5-(Dioctylaminomethyl-quinolin-8-ol in chloroform	-----	-----
Ni (303,306)	Kelex 100 in chloro-benzene	Trioctyl-phosphine oxide	Enhanced % ex-traction
Ga, Al in aqueous NaOH (304)	Kelex 100 in kerosine	-----	-----
Sn from aqueous HCl (305)	Tri-n-octylphosphine oxide in benzene or CCl₄	-----	-----
Th (306)	Aliquat 336 in xylene	Ascorbic acid	-----
Ge in aqueous citric acid (307)	Aliquat 336 in xylene	-----	Eff= 99%
Re,W in aqueous HNO₃ (308)	Adogen 381 in xylene	-----	90% Re[e]
Pd in aqueous HCl (309)	Trioctylamine hydro-chloride in 1-hexene	-----	-----
Al,Ga,In in aqueous oxalic acid (310)	Trioctylamine in CCl₄	-----	-----
Nb,Ta (311)	Trioctylamine in CHCl₃	PAR[f]	-----
Zn,Cd (312)	Zephiramine in Benzene	BMPP[g]	-----
Rh in aqueous HCl (313)	Tetraoctylammonium chloride in toluene	-----	Eff= 98%
Tl in aqueous acetic acid (314)	Trinonyl(octadecyl) ammonium iodide in toluene	-----	-----
Cu from aqueous chloride media (315)	Tri-n-dodecylammonium chloride in toluene	-----	-----
Fe from aqueous sulfate solution (316)	Primene 81R in kerosine, chloroform, or benzene	----	-----
Pt from Cu & Ni in aqueous solution (317)	Alkylmacrocyclic dioxo-tetraamine	----	Eff= 84%

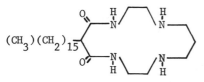

Continued on next page

Table XIV. Continued

Component/Aqueous Conditions (Ref.)	Surfactant/ Organic Solvent	Additives/ Co- extractant	Comments
Alkaline earth metals from water (318)	Polyoxyethylene glycol-4-nonyl-phenyl ether in 1,2-dichloroethane	-----	-----
Lanthanides from aqueous media (319)	Triton X-100 or 405 in dichloroethane	Picrate ion	-----
Au from aqueous HCl (320)	PONPE-7.5 in di-chloroethane	-----	-----
Lanthanides from aqueous media (321)	Phosphate monoester of Triton X-100 in 1,2-dichloroethane	-----	$\alpha = 1.7^h$
Food Dyes (Yellow 4, Red 9, Blue 3) from buffered aqueous solution (322)	Tri-n-octylamine in hexane and $CHCl_3$, CH_2Cl_2, or pentanol	-----	-----
Synthetic dyes from pharmaceutical preparations (323)	Tri-n-octylamine in chloroform	-----	-----
Steroid sulfates from plasma (324)	Benzyltributyl-ammonium chloride or other quaternary ammonium salts in benzene	-----	% Rec = 75 - 100%^{i}
Ligninsulfonates from spent sulfite liquors (325)	Trioctylamine, do-decylamine, or di-octylamine in C_6H_{12}, butanol, or pentanol	-----	-----

^{a}Data for the extraction of components from aqueous (or solid matrices) using surfactant-containing organic solvents.

^{b}Average separation factor between adjacent pairs of analytes extracted.

^{c}Eff refers to the extraction efficiency. ^{d}Refers to the improvement in extraction efficiency. ^{e}Refers to the radiochemical purity after three extraction cycles. ^{f}PAR = 4-(2-pyridylazo)resorcinol.

^{g}BMPP refers to 4-benzoyl-3-methyl-1-phenylpyrozolin-5-one.

^{h}Refers to the mean separation factor of 13 pairs of adjacent lanthanides.

^{i}Refers to the % recovery from plasma samples.

the organic phase (refer to the section on reversed micelles). Additionally, the degree of association of the components of the organic phase can be altered by subtle changes in the experimental conditions (i.e. pH, ionic strength, etc.). Consequently, many apparent divergent results and controversy have been created by the fact that very different aggregation behavior (or no aggregation) can be observed in otherwise very similar extraction systems. Due care must therefore be given to the specific experimental conditions before any comparisons between sets of data can be made.

A simplified pictorial representation of the extraction process is depicted in Figure 5 for alkylammonium extractants (346). At the interface between the aqueous and organic phases are ionized or partially ionized surfactant molecules aligned with their polar head-group in contact with water while their hydrophobic moiety is in contact with the bulk organic solvent. The aqueous phase may contain some surfactant monomer molecules (not shown). The organic phase also contains monomeric surfactant molecules along with the reversed micellar aggregates (in many instances, a distribution of different sized aggregates will exist). Inside the hydrophilic core regions of the reversed micelle can be several solubilized water molecules as well as counterions and solubilized or associated co-extractant molecules (symbolized by AC^+). The ions to be separated (anions or anionic metal complexes), originally in the aqueous phase, are picked up at the interface and transported into the organic phase by the surfactant and/or extractant molecules. Eventually the surfactant-analyte ion pair can interact with the reversed micellar entity and/or the co-extractant molecule present therein. The selectivity of the process is achieved due to differences in the binding of both the co-extractant and the involved ions to the reversed micellar pseudophase and/or to relative differences in the rate of water - extractant ligand exchange reactions. The binding and catalytic micellar effects can be profoundly altered by changes in the solvent (293,322,325,347), pH, ionic strength, nature of the surfactant molecule, etc. (253,332-340).

Recently, some general guidelines have been developed with regard to the potential effects of surfactants upon extraction processes (264,283). Namely, it has been postulated that for an extractant/surfactant to function as a phase transfer catalyst, it must (i) have a greater solubility in the aqueous phase, (ii) preferentially adsorb at the aqueous/organic interface, and (iii) have a greater reactivity (241). Surfactant reversed micellar media can influence an extraction by (i) a metal-extractant concentration effect due to solubilization of the reactant species in the smaller volume element of the reversed micellar system, (ii) their ability to catalyze the water-extractant ligand exchange reactions (241,275, 276), and (iii) mass transfer effects (363). The presence of surfactant phase transfer agents, reversed micelles, or microemulsions can favorably alter interfacial properties and facilitate interphasic transfer of the species to be extracted (349-350,359,363). Thus, as can be seen, the presence of surfactants can influence the rate of extraction which depends upon the kinetics of transfer of the species from the bulk water phase across a water-oil interface

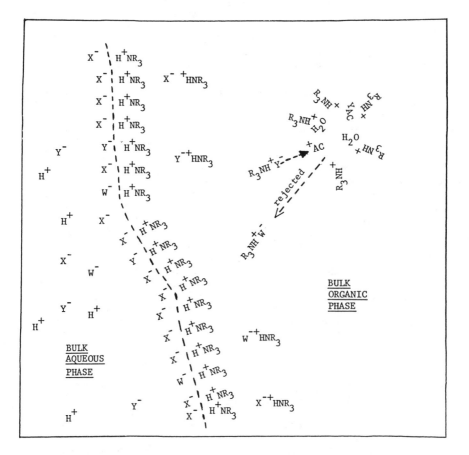

Figure 5. Simplified extraction mechanism for the alkylammonium cationic surfactant system in organic solvents. CA refers to a solubilized co-extractant, X is the counterion of the surfactant, and Y and W refer to the species (anions or anionic metal complexes) to be extracted from the aqueous phase.

to the bulk organic phase (which contains the surfactant/extractant).
Excellent summaries of the current mechanistic thought concerning
micellar interfacial and catalytic effects on extractions are given
in some recent overviews (240,241,253,322,346). Such information
and understanding is likely to aid the design of more efficient
extraction processes.

Lastly, it should be noted that the extracted species in the
organic surfactant-containing phase can be subsequently back-extracted (stripped) back into a suitable aqueous phase. This is
typically done by contacting the loaded organic surfactant media
with an aqueous acidic (or basic) receiving solution at a relatively
high organic/aqueous phase ratio (e.g., 15:1) (263,306,307,323,329,
341-343). Typically, very good recovery of analytes is obtained
along with a good concentration factor. In many instances, a series
of co-extracted ions can be selectively stripped from the organic
phase by contacting to suitable aqueous solutions (306). Procedures
also exist for the regeneration and recycled use of the extracting
organic phase (357). The extracted or back-extracted (stripped)
species are usually detected or determined spectrophotometrically
via use of different-ligand metal complexes (344,345). In the case
of extractions conducted with organophosphorus agents, detection
via use of ICP is possible (348).

A new potentially exciting development in this area of extractions concerns the use of different reversed micellar systems
in countercurrent extractions of different rare earth metals. A
mathematical model was developed in order to help optimize the
different parameters of this new mode of extraction (364). This
should facilitate the further development and utilization of this
approach to metal ion separations.

In addition to metal ion extractions, reversed micelles have
recently been utilized in the separation, recovery, and purification
of biotechnological products (366-371). Namely, Hatton and co-
workers have demonstrated that the controlled solubilization of
proteins and amino acids in reversed micelles can form the basis
of an extractive separation scheme (366-369,377,378,381). Previous
work had indicated that enzymes could be solubilized in organic
solvents containing surfactants (reversed micelles) with retention
of their activity (37,41,365,373,374,382). The recent work of
Hatton and others indicate that the solubilization (and hence degree
of extraction) of a given protein depends upon its structure, especially the ionizable moieties (i.e. pI value) and its size. By
judicious choice of the surfactant charge-type and solvent, and
appropriate manipulation of the experimental conditions (i.e. pH and
ionic strength of the aqueous protein solution), the separation
scientist can control this solubilization process. To date, the
most utilized reversed micelle system has been the AOT (sodium di-2-
ethylhexylsulfosuccinate) anionic surfactant in isooctane one. Thus,
if the aqueous solution pH is such that it is below the pI value of
protein (so that the protein possesses a net positive charge), then
favorable electrostatic interactions between the anionic AOT reversed
micelle and the protein can result in the selective solubilization

of the protein. Consequently, proteins (as well as other substances such as amino acids, etc.) with differing pI values can be separated from each other. This approach using AOT reversed micelles has been successfully utilized to extract ribonuclease-a, cytochrome-c, lysozyme, chymotrypsin, trypsin, rennin, amylase, and amino acids from aqueous solutions (366-369,377). In addition, it was recently shown that the same general approach could be employed to extract directly proteins from solid matrices (i.e. protein powders) (370). For the case of proteins with very large molecular weights, solubilization may not be possible due to size exclusion effects. This factor was apparently responsible for the inability to extract bovine serum albumin from aqueous solution with AOT (367).

Most systems examined to date have employed the AOT anionic reversed micellar system (366-370). In one case, amylase was extracted using trioctylmethylammonium chloride (cationic surfactant) in isooctane (375) while in another, catalase was extracted using a cationic DTAB/octane/hexanol reversed micelle (377). In our own research, we have successfully employed nonionic Igepal CO-530 - CCl_4, cationic CTAB - hexanol, and zwitterionic lecithin - CCl_4 reversed micellar systems in the extraction of some amino acids and proteins (379). The availability of such a pool of different charge-type micellar systems allows one flexibility in the development of such extraction schemes. In fact, preliminary results seem to indicate that better extractions are obtainable in some instances via use of zwitterionic reversed micellar media (379).

In addition to the extraction of extracellular enzymes, the use of reversed micelles can be extended to recover intracellular enzymes from intact bacteria (380). The technique is based on the fact that whole bacterial cells desintegrate readily in such reversed micellar systems and the liberated enzymes are rapidly solubilized by the micelle. This approach should be applicable to other extracellular systems and holds great promise.

The proteins and bioproducts can be easily stripped (i.e. back-extracted) from the reversed micellar organic phase to an aqueous phase. This can be accomplished by merely contacting the reversed micellar loaded organic phase with an aqueous phase of relatively high salt content (e.g. 1.00 M KCl or 0.50 M Na_2SO_4) (368,379). The overall extraction - stripping process has the potential for operation in a continuous mode fashion and should be easily applied on a large scale process-type level (366,371,377,381). Consequently, it appears to be a promising procedure for the bulk purification of biomolecules. In fact, recovery of extracellular alkaline protease from clarified fermentation broths has been achieved (378). Several other studies have examined the nature of reversed micellar formation in fermentation broth media (284,376). Such data should aid in the design of other extraction schemes from such media.

More information on the salient features and factors involved in this exciting new type of bioseparation technique using reversed micellar media is presented in a Chapter by Hatton in this Symposium Volume (377). Future work in this general area should concentrate on the mechanistic aspects as well as on the continued development of schemes using different charge-type reversed micellar systems. Also, use of functionalized surfactants as the reversed micellar formers (8) tailored for specific components of mixtures should be synthesized and evaluated. Further development of specific affinity ligand co-surfactants for such tailored extractions should also lead to selective separation and purification of biomaterials (377).

Applications of Aqueous Surfactant Solutions as a Selective Solubilization Extraction Medium. (1) Extractions Utilizing Aqueous Normal Micelles. Selective solubilization in aqueous surfactant micellar media has been employed in the extractive isolation and purification (often only partial) of various biochemical compounds and membrane components (70,98,103-112,237,238,383-392, 402-404). For example, such biomaterials as D-glucosidase, human tissue factors, lysosomal acid lipase, lactoferrin, uricase, neuromembrane microsomes, cholesterol oxidase, and malarial acid endopeptidase have been extracted and purified via use of Tween-20, Tween-80, Triton X-100, Triton N-101, NaDC, NaTC, NaLS, and octyl glucoside aqueous micellar media among others (405-414). A peculiar feature observed in the solubilization of membranes in surfactant solutions is the selectivity with which different components are released from the membranes (238). This selectivity can be fine "tuned" by judicious variation of such parameters as solution pH, ionic strength, concentration and nature of the surfactant [i.e. its charge-type and HLB (hydrophile-lipophile balance)]. Such extraction procedures have proven to be extremely useful and valuable in the partial purification of active membrane complexes, membrane lipids, and membrane proteins among other components (383-392). For instance, cytochrome o and d oxidase have been extracted and purified from $\underline{E}$. coli by use of zwitterionic sulfobetaines or nonionic octylglucoside normal micellar systems (415,416).

Although a few general guidelines exist, the design of such mentioned extraction schemes is still a trial-and-error proposition. Consequently, more basic information on the nature of the lipid - protein - surfactant interactions is still required. It should also be noted that in most instances, the micellar "extraction" step is merely the prelude to further fractionation (usually by electro-phoretic, column or hydrophobic chromatographic techniques) and purification of the desired biological components (402-404).

Normal aqueous micellar media can also be employed to extract and purify components from solid matrices. Proteins have been extracted from wheat kernals using aqueous NaLS (399). This same sur-factant system has been employed in an improved method for the extraction of filth from cheese (417). In another application, aqueous solutions of Brij-35 micelles have been employed to extract components (i.e. vanillin and ethylvanillin) from smoking tobacco (106). In a similar manner, various phenolic compounds have been extracted from herbal/plant leaves using nonionic Triton X-100, Brij-35, or octyl glucoside (OG) (393). In both of these latter examples, the indicated compounds could be identified and quantitated by reversed phase HPLC using as mobile phase the same micellar solutions (refer

48   ORDERED MEDIA IN CHEMICAL SEPARATIONS

to previous section on micellar LC) as were employed in the extraction step. The recent development of micellar mobile phases
should provide further impetus for the development and use of
aqueous surfactant solutions in the extraction of other nonpolar and
biological substances from various matrices. The use of aqueous
micellar media as opposed to the usual organic solvents in such
extraction steps offers advantages in terms of safety (diminished
toxicity, flammability), easy waste disposal, cost, as well as en-
sures compatibility of the extraction "solvent" with chromatographic
micellar mobile phases.

Aqueous micellar systems have also been utilized to extract
organic components from hydrocarbon matrices (394-398,400,401).
The basis of the extraction process lies in the ability of the
aqueous micellar media to exhibit differential solubilizing rates
and capacities with respect to a series of solubilizates. Table XV
summarizes some of the data along with the separation factors that
were achieved. The industrial potential of the approach is demonstrated by the extractive recovery of phenols from carbolic oil
(Table XV) (397,400). Kinetic and thermodynamic models of the extraction process have been formulated which attempt to relate the
selectivity, extent, and rate of solubilization to the structure
and properties of the surfactant micelle as well as to the nature of
the solute molecules (394,395,397). In general, "a solubilizate with
the smaller molecular volume and some polarity is usually preferentially solubilized" by a particular micelle (401). In such extraction schemes, the adsorption processes of the surfactant at the
aqueous - organic interface seem to provide the main resistance to
mass transfer (396). It should be noted that microemulsions form in
some of these extraction systems (395,401). It seems that this
approach is an excellent means of removing aromatic hydrocarbons from
mixtures of aliphatic and aromatic hydrocarbons (394,398). This is
one of the few areas in which the current work seems to be more
mechanistic than practical in nature. This general area of separation science appears to hold great potential for development of some
practical large-scale bulk extraction schemes for industrial processes.

    (2) Extractions Based on the Phase Separation Behavior of Aq-
ueous Micellar Solutions. The extraction and concentration of components in an aqueous mixture can sometimes be effected via use of
appropriate surfactant systems that are capable of undergoing a
phase separation as a result of altered conditions (i.e. temperature
or pressure changes, added salts or other species, etc.). Two
general types of such surfactant extraction systems will be described: (i) those based on the cloud point phenomenon and (ii)
those based on coacervation formation.

    (i) Cloud Point Separations. Aqueous solutions of nonionic
surfactants (442), such as n-alkylsulfinyl alcohols, alkylmethyl-
sulfone-di-imines, dimethylalkylphosphine oxides, and most commonly,
alkyl(or aryl)polyoxyethylene ethers, exhibit the so-called cloud
point phenomena (418-428). That is, upon heating the isotropic
micellar solution, a critical temperature is eventually reached at

TABLE XV. Summary of Separation Factors for the Extraction of
Organic Substances from Nonpolar Matrices via Use of
Normal Aqueous Micellar Systems

Original Mixture (Ref)	Micellar System	Selectivity Factor
Benzene/Cyclohexane (401)	CPC^a	$1.5 - 1.8^b$
Benzene/Hexane (394,401)	DTAC	2 - 7
	Aerosol AY	2 - 3
	NaLS	7
	CPC	7 -10
	OG	7 -10
Cyclohexane/Hexane (401)	CPC	1.25
Dodecane (395)	$C_{12}E_4{}^c$	----
Benzene/Cyclohexane (398)	DTAB	----
Phenols in Carbolic Oil (397, 400)	NaLS	$----^d$
	Sodium Oleate	4.0^d
	NaDBS	$----^d$
	Sodium Laurate	5.2^d
	Potassium palmitate	3.6^d

[a]Refer to Table I for details of micellar structure and parameters.

[b]Separation factor defined as $\alpha = {}^x_{FE}{}^x_{oR}/{}^x_{oE}{}^x_{FR}$, where x = concentration (in kg/kg); F = benzene, cyclohexane, or phenols; o = hexane, cyclohexane, or oil; R = raffinate; and E = extract (397).

[c]Dodecylpolyoxyethylene(4)glycol monoether (CMC = 6.2×10^{-5} M).

[d]Limiting values. In general, the selectivity of the extraction process increases with surfactant concentration until a plateau is achieved.

which the solution suddenly becomes turbid (cloud point) due to the
diminished solubility of the surfactant in water. After some time
interval, demixing into two transparent liquid phases occurs; i.e.
a surfactant-rich phase in equilibrium with almost pure water (8,
418,425). Figure 6 shows a typical temperature - concentration
diagram which defines the cloud point at each surfactant concentra-
level (421,425). This coexistence curve exhibits a lower consolute
point (critical point). The temperature and concentration at which
the minimum occur are called the critical temperature (T_c) and
critical concentration (c_c), respectively. Table XVI summarizes
cloud point and CMC data for some representative aqueous nonionic
micellar surfactants (8,425). It is important to point out that the
presence of additives (salts, organic species, alcohols, etc.) can
profoundly alter (either raise or lower) the reported values. The
mechanistic details concerning the dynamics, cause, and nature of
the clouding phenomenon in relation to the aggregation behavior is
still not completely understood and the subject of some controversy
(422-428).

Based upon the use of nonionic surfactant systems and their
cloud point phase separation behavior, several simple, practical,
and efficient extraction methods have been proposed for the separation, concentration, and/or purification of a variety of substances
including metal ions, proteins, and organic substances (429-441,
443,444). The use of nonionic micelles in this regard was first
described and pioneered by Watanabe and co-workers who applied the
approach to the separation and enrichment of metal ions (as metal
chelates) (429-435). That is, metal ions in solution were converted
to sparingly water soluble metal chelates which were then solubilized by addition of nonionic surfactant micelles subsequent to
separation by the cloud point technique. Table XVII summarizes
data available in the literature demonstrating the potential of the
method for the separation of metal ions. As can be seen, factors
of up to forty have been reported for the concentration effect of
the separated metals.

The extraction procedure typically consists of the following
steps: (1) addition of 0.5 - 1.0 mL of an appropriate concentrated
solution of the nonionic surfactant to 50 or 100 mL of the metal-
containing sample buffered to an appropriate pH in the presence of
a suitable chelating agent and/or masking agent, (2) mixing, (3)
heating (if necessary) of the solution to a temperature in excess
of the surfactant's cloud point, (4) waiting interval which facilitates the solution phase separation (centrifugation can speed up
this step), and (5) physically removing and analyzing the desired
enriched surfactant phase containing the metal chelate (1,429-431).
The most recent advance in this area has concerned the synthesis
and use of functionalized surfactants having chelating moieties in
conjunction with various nonionic solubilizing surfactants (1,437).
Such an approach results in novel and more selective protocols for
a variety of analytes.

In addition to concentration of metal ions, biological materials and environmentally important organic compounds have also been

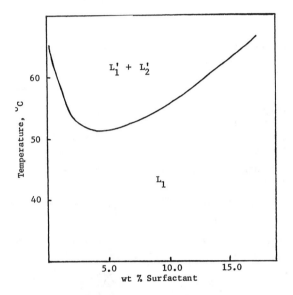

Figure 6. Phase diagram for a typical nonionic surfactant. L_1 region refers to an isotropic amphiphile solution whereas L_1' and L_2' indicates the two co-existing isotropic phases.

TABLE XVI. Summary of Structure of Some Selected Nonionic Surfactants and their CMC and Cloud Point Values[a]

Surfactant	CMC, M	Cloud Point[b], °C	Ref.
Triton X-100	3.0×10^{-4}	64	393, 430
Triton X-114	---	23	8, 393
Brij-96 (polyoxyethylene ether of oleyl alcohol)	9.2×10^{-5}	54	427
Brominated Brij-96	1.0×10^{-4}	60	393
Brominated Tween-80	1.85×10^{-5}	90	393
PONPE-7.5 (polyoxyethylenenonylphenyl ether)	8.5×10^{-5}	1 - 7	429-431
IGEPAL SERIES: $R-C_6H_4O(C_2H_4O)_{n-1}CH_2CH_2OH$			
CO-620 (R = C_8H_{17}; n = 7)	---	22	393
CO-610 (R = C_9H_{19}; n = 7.5)	8.0×10^{-5}	26	393
ALKYLPOLYOXYETHYLENEGLYCOL MONOETHER SERIES:			
C_8E_3[c]	7.8×10^{-3}	8	11, 425, 439
C_8E_4	8.5×10^{-3}	40	
$C_{10}E_3$	---	0	
$C_{12}E_3$	5.8×10^{-5}	0	
$C_{12}E_4$	---	4	
$C_{12}E_5$	6.5×10^{-5}	31	
$C_{16}E_3$	---	20	

[a]Values given for 1.0% aqueous surfactant solutions. [b]Value in pure water alone in absence of any additives.
[c]In this shorthand notation, the subscript after C refers to the number of carbons of the n-alkyl group while the subscript after E refers to the number of oxyethylene groups present.

TABLE XVII. Summary of the Reported Extraction of Metal Ions as
Metal Chelates by the Phase Separation Behavior of
Nonionic Surfactants

Metal Ion	Chelating Agent	Surfactant	Concentration Factor	Ref.
Zn	PAN[a]	PONPE-7.5	40	431, 433
Ni	TAN[b]	TX-100	30	432
Ni	PAN	OP	15 - 25	436
Zn	PAMP[c]	PONPE-7.5	---	430
Au (in HCl)	----	PONPE-7.5	($\geqq$ 95% recovery)	443
Cd,Ni	PAMP	PONPE-7.5	---	434
Fe	HAHBA[d]	TX-100/BL 4.2[e]	(Extn. Eff. 94%)	437

[a]PAN refers to 1-(2-pyridylazo)-2-naphthol.

[b]TAN refers to 1-(2-thiazolyazo)-2-naphthol.

[c]PAMP refers to 2-(2-pyridylazo)-5-methylphenol.

[d]HAHBA refers to 4-Heptylamido-2-hydroxybenzoic acid.

[e]BL 4.2 refers to polyoxyethylene(4.2)dodecanol.

extracted using the described general approach. For instance,
Bordier has reported the separation of peripheral membrane proteins
from intrinsic membrane proteins by use of Triton X-114 (439).
When the membrane dispersion in TX-114 was heated, phase separation
occurred and the integral membrane proteins were concentrated in the
surfactant rich phase while the hydrophilic peripheral proteins remained in the aqueous phase. Recently, a report stated that such
extractions using polyglycol ethers is simple and should be industrially practicable for the separation, isolation, and purification
of may types of hydrophobic proteins (444). Additionally, Kippenberger et. al. (440) have reported that various organic compounds
of environmental concern can be essentially quantitatively concentrated in the surfactant rich phase using such described nonionic
extraction scheme. For example, a pesticide test mix containing
DDT, methoxychlor, endrin aldehyde, endrin, chloradane, endosultan,
lindane, aldrin, BHC, and chloropyrifos was successfully extracted
from water using PONPE-7.5. Similar results have been obtained with
phenolic, polynuclear aromatic hydrocarbon, and aniline test mixes.
The approach shows very promising results in terms of concentrating
these materials piror to gas or liquid chromatographic analysis
(393).

The percent recovery of the extracted substances in these systems has been found to be dependent on the solution pH (for ionizable analytes), presence of additives, nature of surfactant, etc.
A quantitative description of the distribution coefficient variation
with solution pH has been reported (435) (equation 9):

$$D = \frac{P_1[H^+]/K_{a1} + P_2 + P_2P_3/[H^+]}{1 + [H^+]/K_{a1} + K_{a2}/[H^+]}$$ (eq. 9)

where D is the distribution coefficient; P_1, P_2, and P_3 are the
partition coefficients of the positively charged, neutral and negatively charged species, respectively; K_{1a} and K_{2a} are the acid
dissociation constants; and $[H^+]$ is the hydrogen ion concentration.
The development of this quantitative description is important in
that it allows one to optimize the analytical conditions so as to
maximize the extraction efficiency. The distribution coefficients
determined using the cloud point extraction method compare favorably
with those obtained using more conventional organic-based extraction
schemes. The recovery of metal chelates and organic substances
typically range from 70 - 100% (429-431,440).

The advantages cited for the described nonionic micellar cloud
point extraction schemes include the following: (1) ability to
concentrate a variety of analytes (with concentration factors of
10-75), (2) safety and cost benefits (i.e. the use of small amounts
of nonionic surfactant as an extraction solvent obviates the need to
handle the usually large volumes of organic solvent required in
traditional liquid-liquid extractions; so that the volatility,
flammability, and cost are reduced), (3) easy disposal of the nonionic surfactant extraction solvent (i.e. the nonionic surfactant
solution is reportedly easily burned in the presence of waste

acetone or ethanol), (4) the surfactant rich extraction phase is
compatible with micellar mobile phases in TLC or HPLC, (5) possibility of enhanced detection is possible in the surfactant rich
phase compared to that possible in bulk water or organic solvents
(enhanced detection will be described in a later section of this
review), and (6) ability to treat wastewater solutions in water
purification/monitoring schemes (429,431,440). For more details on
this separation technique, the interested reader is referred to a
review by Watanabe (429) as well as to Chapter by Pramauro, Minero,
and Pelizzetti in this Symposium Volume (437).

Future work in this area should focus on further development of
novel extraction schemes that exploit one or more of the cited
advantages of the nonionic cloud point method. It is worth noting
that certain ionic, zwitterionic, microemulsion, and polymeric
solutions also have critical consolution points (425,441). There
appear to be no examples of the utilization of such media in extractions to date. Consequently, the use of some of these other
systems could lead to additional useful concentration methods;
especially in view of the fact that electrostatic interactions with
analyte molecules is possible in such media whereas they are not in
the nonionic surfactant systems. The use of the cloud point event
should also be useful in that it allows for enhanced thermal lensing
methods of detection.

(ii) Separations Based on Coacervation. Some ionic surfactant
solutions are also capable of separating into two liquid layers under
appropriate conditions. This type of liquid-liquid phase separation
has been termed coacervation. Coacervation can occur when a species
(which contains a charge opposite to that of the surfactant) is added to an aqueous solution of an ionic micelle (such as a cationic
quaternary ammonium surfactant). Coacervation in such solutions be-
gins with the micelles aggregating to form submicroscopic "clusters"
which can coalesce to form microscopic droplets. On further coales-
cence, these droplets can separate into a continuous surfactant-rich
phase. The two phases thus formed are well defined. With further
addition of electrolyte, the surfactant-rich phase can precipitate
or flucculate (445). In addition to cationic and anionic surfactants
(445-452), aqueous solutions of proteins, synthetic polymers, and
microemulsions have been reported to exhibit coacervation behavior
(445). Most literature reports on the topic have been conducted on
systems in which the surfactant concentration were in excess of the
CMC so that micelles were present. In such media, a certain amount
of added electrolyte (critical electrolyte concentration, CEC) is
required to induce coacervation. For example, at 30° C in the
presence of 0.15 M ammonium thiocyanate, CTAB forms a coacervate
system (445). In many coacervate systems, the two phases are well
defined with one rich in surfactant and the other essentially pure
water. Thus in theory, separations identical to those just outlined employing the nonionic cloud point phenomenon (a temperature
induced coacervate) should be possible.

A cursory review of the literature reveals that there are only
two apparent applications of coacervate formation in extractive

separation science. One report mentioned that "the analogy between
these (coacervate) systems (i.e. composed of tetraalkylammonium
halides) and anionic ion exchange resins is shown for uranium(VI)
extractions" (451). In the other application, the coacervation process was utilized to recover (at better than 90% efficiency) and
purify the surfactant dodecyldimethylbenzylammonium bromide from a
froth flotation system (452). One possible disadvantage of coacervate systems (compared to the nonionic cloud point extractions) is
that the liquid-liquid two-phase state is difficult to experimentally
access. Thus, since most ionic surfactants are solids, the micelle-
rich phase can precipitate necessitating a redissolution step
prior to quantitation or further work-up of the extracted species.
However, this entire coacervate research area is wide open and
further work will undoubtly lead to many novel applications in
separation science. An article by Dubin and co-workers in this
Symposium Volume gives more details on coacervation as it pertains
to protein - polyelectrolyte systems (453).

Other Applications of Surfactant Assemblies in Separation Science.
      Surfactants in Membrane Processes. (1) Micelle-Enhanced Ultra-
filtration (MEUF). In this newly described technique, micelle-
forming surfactants are added to the aqueous phase in question so
that micelles form in solution. Any hydrophobic organic molecules
originally present (or metal ions/metal chelates that can bind the
micelle) will partition to the micellar pseudophase. This solution
is then passed through an ultrafiltration membrane whose pore size
is small enough (i.e. mol. wt. cutoff $\leq$ 20,000) to prevent micelles
from passing through it. Consequently, the micellar bound organic
molecules (or metals) are concentrated in the retentate while the
permeate contains only the unsolubilized solute and the surfactant
monomer (both present at relatively low concentrations). More details on the experimental considerations and salient features of
this new separation technique are given in an excellent overview
article (454) as well as in a Chapter by Smith, Christian, Tucker,
and Scamehorn in this Symposium Volume (455).

      In terms of published practical applications, MEUF has been
successfully employed to remove multivalent metal cations (such as
Cu^{2+}) from wastewater (456), for removal of n-alcohols (457) as well
as other organic substances (benzene, etc.) (454,455) from aqueous
streams. The general procedure seems to offer an attractive means
for purification of water streams containing organic substances or
heavy metals. MEUF may prove to be useful in industrial applications
as well (455).

      In connection with MEUF efficiency studies, the technique of
"semi-equilibrium dialysis" (SED) was developed. SED utilizes
ordinary commercial equilibrium dialysis apparatus and membranes.
The method allows for determination of the concentration of the
surfactant passing through the membrane at varying times (454).
Via use of the SED technique, it is possible to determine solute
solubilization equilibrium constants (activity coefficients of
micellar bound solutes) as well as the surfactant concentration on
both sides of the dialysis membrane among other parameters (458-460).

(2) <u>Liquid</u> <u>Surfactant</u> (or <u>Liquid</u> <u>Surfactant</u> <u>Supported</u>)
<u>Membranes</u>. Since their first description in the early 1960's (<u>461</u>),
the use of liquid surfactant (or supported or immobilized) membranes
has become an increasingly popular procedure for the isolation and
purification of a variety of analytes. A limited summary of some of
the different systems reported is given in Table XVIII. As can be
seen, aqueous surfactant solutions can serve as liquid membranes for
the separation of components in organic solvents whereas organic
surfactant-containing solutions can be utilized to separate components in aqueous media. Figure 7 gives a simple schematic of a
liquid membrane containing an organic surfactant solution as the
active component. In this example, the source phase would be an
aqueous solution containing the components to be separated. The
surfactant dispersed in the organic solvent serves as the transport
vehicle and can be present in the form of monomers and thus act as
an ion pair or phase transfer agent; or the surfactant can be aggregated in the form of reversed micelles or microemulsions. The
experimental conditions dictate the form that the surfactant will be
present as. Likewise, if the analyte components are in an organic
solvent, then the liquid membrane will consist of the surfactant
dispersed in an aqueous phase with the surfactant present in either
monomer or micellar form. Regardless of the system, due to the
selective interactions of the different analyte components with the
"active form" of the surfactant which constitutes the membrane and
due to mass transfer effects, differences in the permeation rates
are possible. This serves as the basis for separation using this
technique.

Since its introduction by Li (<u>461</u>), liquid surfactant membranes
have been utilized for separation of hydrocarbons, recovery of
amines, phenols, and organic acids from waste water, as well as the
separation and recovery of metal ions. Table XVIII gives data on
some of these selected separation systems. It should be mentioned
that many of the guidelines and knowledge gained from the bulk
extraction systems involving reversed micelles and aqueous normal
micelles described in the previous section facilitate design of
successful liquid membrane devices. This is not surprising since
the membrane-based separation merely combines the extraction step
with the back-extraction stripping step. Numerous recent reports
have concerned model studies on the membrane transfer processes
(<u>474-481</u>). Such data is useful and should enable the separation
scientist to develop better liquid surfactant membranes for desired
applications. An exciting new advance in this field is the use of
AOT reversed micellar media as the liquid surfactant membrane for
isolation and purification of proteins/enzymes that has been described by Armstrong and co-workers (<u>473</u>). In our work, we have employed chiral surfactants and achieved partial chiral resolution of
some amino acids (as their dansylated derivatives) (<u>393</u>).   Future
work should concentrate on further development and expansion of
applications of liquid surfactant membranes in the purification of
biomaterials and in optical resolutions. The use of such membranes
should allow for purifications at preparatory and process levels
(<u>473</u>).

TABLE XVIII. Limited Summary of Separations Based on the Use of
              Liquid (or Liquid Supported) Surfactant Membranes

Analyte Separated/ Purified (Ref.)	Liquid Surfactant Membrane System	Comments
Cu(II) (462)	HDEHP/cyclohexane reversed micelle	% Recovery ~ 95 %
U(VI) (463)	Trioctylphosphine oxide supported on a microporous polymer	% Recovery ≦ 99 %
Cu(II) & Cr(III) (464)	LIX 64, HDEHP, or Aliquats on solid supports	-------
U(VI) (465,468)	Kelex 100, HDEHP, or trioctylphosphine oxide in organic solvents	% Recovery ≧ 90 %
Ge from Zn electrolytic solutions (466)	KELEX 100 on inert macroporous support	-------
Lanthanides & Actinides (469)	CMPO in decalin absorbed on thin polypropylene supports	applied to nuclear wastes
Cu(II) (467)	Span 80, LIX 64 in kerosine microemulsion	-------
Hydrocarbons (472)	Liquid film of aqueous surfactants (micelles)	-------
Rare earth metals (471)	Span in kerosine	% Recovery ≧ 99 %
Positional isomers (o-, m-, p-xylenes) (470)	Aqueous surfactant solutions	-------
Proteins (chymotrypsin, albumin, globulins, etc.) (473)	AOT/isooctane reversed micelles	-------
Enantiomers (dansyl amino acids) (393)	Cationic polyelectrolyte quarternary ammonium surfactants in aqueous media	Partial resolution obtained

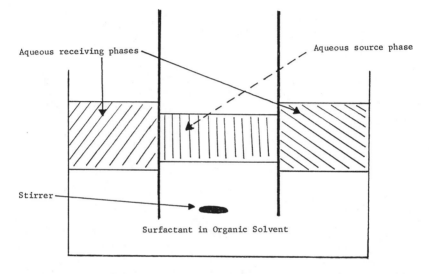

Figure 7. Simplified representation of a bulk liquid surfactant membrane in which the membrane is composed of a surfactant-containing organic solvent.

Micellar-Enhanced Detection in Separation Science. In several
instances, the use of appropriate surfactant media allows for enhanced and/or new modes for chromatographic detection. Most of the
work has been in the area of spectroscopy. For example, it is well
known that the presence of suitable surfactant micelles can significantly increase the absorbance of metal complexes (4,8,345).
This effect has been successfully employed to improve the LC detection of metal ions as their metal complexes (496,497,499). Recently,
it has also been demonstrated that metal ions can be detected by
direct-current argon plasma emission spectroscopy after LC separation
with micellar mobile phases (490).

Likewise, the luminescence properties of many analytes can be
altered in the presence of surfactant aggregates (4,7,8). Consequently, addition of micelle-forming surfactants (present either in the LC
mobile phase or added post-column) can improve the sensitivity of
fluorimetric LC detectors (49,482). Micellar spray reagents have
been utilized to enhance the fluorescence densitometric detection
of dansylamino acids or polycyclic aromatic hydrocarbons (483). The
effect was observed for TLC performed on cellulose or polyamide
stationary phases with the micellar spray reagent being either CTAC,
SB-12, or NaC (483). More recently, use of nonionic Triton X-100
has been found to improve the HPLC detection of morphine by fluorescence determination after post-column derivatization (486) as well as
improve the N-chlorination procedure for the detection of amines,
amides, and related compounds on thin-layer chromatograms (488).

More importantly, the use of heavy metal anionic micellar media
has been shown to allow for observation of analytically useful room-
temperature liquid phosphorescence (RTLP) (7,484,487). There are
several examples in which phosphorescence has been employed as a
LC detector with the required micellar assembly being present as
part of the LC mobile phase (482) or added post column (485). More
recently, metal ions have been determined in a coacervate scum by
utilizing the micellar-stabilized RTLP approach (498). Thus, the
future should see further development in RTLP detection of metal
ions in separation science applications.

Typically, in gradient elution liquid chromatography, electrochemical detection has been difficult due to base-line shifts that
result as a consequence of the altered mobile phase composition.
However, a unique property of micelles allows for much improved
compatibility of gradients (i.e. gradient in terms of micellar
concentration or variation of small amount of additive such as
pentanol) with electrochemical detectors. This has been demonstrated
by the separation and electrochemical detection of phenols using
micellar gradient LC (488). A surfactant (apparently non-micellar)
gradient elution with electrochemical detection has also been successfully applied for the assay of some thyroid hormones by LC (491).

A micellar fluorescence quenching spray reagent has been utilized to allow for Raman detection of a series of metalloporphyrins
separated by TLC (492). Previously, such detection was not possible
due to the strong intereference of fluorescence which obscured the
weak Raman lines (495). Consequently, the utilization of this

approach allows for an extension of the types of compounds that are amenable to detection by Raman spectroscopy in HPLC or TLC.

Other Useful Applications. It is well known that there are many other important applications of surfactants and organized surfactant assemblies in separation science. Many specific separation processes such as secondary and tertiary oil recovery (500-502), tar sand extraction (503), gas scrubbing and purification (504) and different electrophoretic techniques utilize surface active agents (505). However, space limitations and the existence of several recent review articles preclude further discussion of these applications in this particular overview.

Experimental Considerations in Use of Surfactant Media in Separations

For the most part, the commercially available surfactants may be utilized as received without further treatment. Various techniques for the analysis and purification of surfactants are reported in literature (12,16,20,26,506-508). In those applications requiring degassed surfactant solutions in the separations scheme, it is easiest to first degas the bulk solvent prior to addition of the surfactant in order to minimize foaming problems.

The most significant problem with the utilization of surfactant media in different separation schemes (particularly those at the preparative or process scales) concerns the recovery of the analyte from the surfactant media and subsequent recovery of the surfactant for re-use. Attempts to use extraction schemes with conventional organic solvents typically results in troublesome emulsion formation during the recovery steps. There are, however, several means available by which analytes can be recovered free of surfactant. These include the following: (1) Several quick, gentle methods for the recovery of some analytes (usually proteins) from surfactant media (i.e. micellar NaLS, Triton X-100, CHAPS, deoxycholate, Brij-35) via use of column chromatography have been developed (509-515). Most of the stationary phase materials for this approach are available commercially (510,513).

(2) A second general approach for the isolation of analytes from surfactant micellar media involves an extraction-precipitation technique. Namely, the organic analyte can be extracted into an organic solvent (such as hexane) as the surfactant is precipitated by the slow addition of appropriate salt (calcium chloride for NaLS or NaL; sodium perchlorate for CTAB or CTAC surfactants) to a rapidly stirred mixture of the micellar solution and hexane (515). This approach has worked well in the recovery of products from organic reactions conducted in aqeuous micellar media (515) and should be useful in some situations encountered in separation science. Also, the precipitated surfactant can be ion-exchanged back to its original form and re-used again.

(3) The third approach, developed in organic chemistry, involves the use of so-called "destructible" surfactants (516-519). Destructible surfactants possess a labile bond that can be hydro-

lyzed under appropriate pH conditions (either acidic or basic depending upon the particular surfactant employed) to form non-sur-
factant hydrolysis products (516). Consequently, these surfactants
could be employed in the particular separation science application,
and subsequently hydrolyzed during the workup which converts them to
nonsurfactant products, thereby eliminating any emulsion problem and
facilitating use of straightforward analyte recovery procedures (i.e.
distillation, extraction, or chromatography). A series of such
destructible surfactants with a wide range of stability/lability
characteristics (with respect to the pH at which they hydrolyze) has
been described and characterized (516-518). An overview on the preparation and properties of these type surfactants is available (519).
Hence, these type surfactants should find use in many separation
science schemes. It is important to note that these destructible
type surfactants can form micellar assemblies and have the same
general properties as those described previously for the more common
surfactants employed to form micelles (516-519).

     (4) Lastly, it may be possible to recover some analytes from
the micellar/surfactant media by distillation. Several patent reports claim that materials (mostly essential or edible oils) can be
recovered from highly concentrated micelles in this manner (520,521).
The abstracts are too vague to judge the relative merit of this
procedure or whether it is applicable to actual separation science
problems. Further work is obviously required in this area.

Future Applications and Areas for Research Opportunities

     In addition to the specific areas already mentioned in this over-
view, the author believes that there are several other exciting possibilities for advancement of the utilization of surfactant media in
separation science. One fascinating potential application is the use
of surfactants in supercritical extraction systems or as the mobile
phase in supercritical fluid chromatography (SFC). For instance,
carbon dioxide gives a supercritical fluid under appropriate conditions (522) and has been utilized in extractions and chromatographic separations. However, it is limited in the polarity of
compounds extracted or separated. Thus, if reversed micelles can
exist at supercritical conditions, then it would be possible to extend the range of application of these techniques to more polar
substances. There have been preliminary reports of the study of the
aggregates under supercritical fluid conditions (523) as well as the
use of SFC with reversed micellar (524,525) or other surfactant
media (524,526). In addition to the enhanced separations using
micellar supercritical fluids, such systems would be ideally suited
for the observation of micellar-improved thermal lensing detection
(244,379). There will no doubt be significant advances in this area
of research in the future which will benefit separation scientists.

     The use of other novel surfactant systems, such as chiral,
fluoro, or functional surfactants should aid in development of more
selective procedures in many areas of separation science due to the
specific chiral, fluorine, or binding interactions possible. We have
recently managed partial optical resolution using chiral mixed micel-

lar mobile phases in LC (244). The use of functional co-surfactants
or functionalized surfactant ordered media as already mentioned in
previous sections of this overview (and in two other Chapters of
this Symposium Volume (377,437) will continue to develop and ultimately be very useful in many areas of separation science. Lastly, the utilization of surfactant media in field flow fractionation
and countercurrent chromatography will extend the usefullness of
these two techniques (527).

ACKNOWLEDGMENTS

The author wishes to thank D. W. Armstrong, S. G. Weber, L. Guiney,
and two anonymous reviewers for their helpful comments with regard
to the preparation of this manuscript. The support of our own work
mentioned in this overview by the National Science Foundation (CHE-
8215508), The Petroleum Research Fund, and the North Carolina Biotechnology Center is gratefully acknowledged. This paper was presented in part at the Workshop on Colloids and Surfactants: Fundamentals and Applications which was held in Belgirate, Italy, June
10, 1986.

Literature Cited

1. Armstrong, D. W. Sep. Purif. Methods 1985, 14, 212.
2. Armstrong, D. W. Am. Laboratory 1981, 13, 14.
3. Armstrong, D. W. In "Solution Behavior of Surfactants";
   Mittal, K. L.; Fendler, E. J., Eds.; Plenum Press: New York,
   1982; Vol. 2, p. 1273.
4. Hinze, W. L. In "Solution Chemistry of Surfactants"; Mittal,
   K. L., Ed.; Plenum Press: New York, 1979; Vol. I, pp. 79 -
   127.
5. Hinze, W. L. In "Colloids and Surfactants: Fundamentals and
   Applications"; Barni, E.; Pelizzetti, E., Eds.; Ann. Chim.,
   Societa Chimica Italiana: Rome, 1987; in press.
6. Saitoh, K. Bunseki (Japan) 1982, 8, 594.
7. Cline Love, L. J.; Habarta, J. G.; Dorsey, J. G. Anal. Chem.
   1984, 56, 1132A.
8. Pelizzetti, E.; Pramauro, E. Anal. Chim. Acta 1985, 169, 1.
9. Fendler, J. H. J. Membrane Sci. 1986, in press.
10. Zeng, H. Huaxue Tongbao 1986, (3), 22 - 26; Chem. Abstr.
    1986, 104: 230999c.
11. Wennerstrom, H.; Lindman, B. Phys. Reports 1979, 52, 1 and
    references therein.
12. Fendler, J. H.; Fendler, E. J. "Catalysis in Micellar and
    Macromolecular Systems"; Academic Press: New York, 1975.
13. Rosen, M. J. "Surfactants and Interfacial Phenomena";
    Wiley-Interscience: New York, 1978.

14. Tanford, C. "The Hydrophobic Effect: Formation of Micelles and Biological Membranes"; John Wiley & Sons: New York, 1973.
15. Fendler, J. H. "Membrane Mimetic Chemistry"; Wiley-Interscience: New York, 1982.
16. Attwood, D.; Florence, A. T. "Surfactant Systems"; Chapman & Hall: London, 1983.
17. Menger, F. M. Acc. Chem. Res. 1979, 12, 111.
18. Goodman, J. F.; Walker, T. In "Colloid Science"; Everett, D. H., Ed.; The Chemical Society: London, 1979; Vol. 3, Ch. 5, pp. 230-252.
19. Mukerjee, P.; Mysels, K. "Critical Micelle Concentrations of Aqueous Surfactant Systems"; National Standards Reference Data Series, Vol. 36, National Bureau of Standards: Washington, D.C., 1971.
20. Small, D. M. In "The Bile Acids"; Nair, P. P.; Kritchevsky, D.; Eds.; Plenum Press: New York, 1971; Chapter 8.
21. Oakenfull, D. In "Aggregation Processes in Solution"; Wyn-Jones, E.; Gormally, J., Eds.; Elsevier: New York, 1983; Chapter 5.
22. "Chem Sources - U.S.A."; Baker, M. J.; Gandenberger, C. L.; Krohn, C. L.; Krohn, D. A.; Merz, J. B.; Tedeschi, M. J., Eds.; Directories Publishing Co.: Clemson, SC, 1986.
23. Hofmann, A. F.; Mekhjian, H. S. In "The Bile Acids"; Nair, P. P.; Kritchevsky, D., Eds.; Plenum Press: New York, 1973; Vol. 2, p. 119.
24. Kratohvil, J. P.; Hsu, W. P.; Kavok, D. I. Langmuir 1986, 2, 256.
25. Roe, J. M.; Barry, B. W. J. Colloid Interface Sci. 1985, 107, 398.
26. Ash, M.; Ash, I. "Encyclopedia of Surfactants"; Chemical Publishing Co., Inc.: New York, 1981; Volumes I-III; 1985; Volume IV.
27. Magid, L. In "Solution Chemistry of Surfactants"; Mittal, K. L., Ed.; Plenum Press: New York, 1979; Vol. I, p. 427.
28. Ionescu, L. G.; Romanesco, L. S.; Nome, F. In "Surfactants in Solution"; Mittal, K. L.; Lindman, B., Eds.; Plenum Press: New York, 1984; Vol. 2, p. 789; "Abstracts of Papers", Sixth International Symposium on Surfactants in Solution, August, 1986, Abstr. No. MS14R, p. 14.
29. Bunton, C. A.; Gan, L. H.; Hamed, F. H.; Moffatt, J. R. J. Phys. Chem. 1983, 87, 336.
30. Evans, D. F.; Yamauchi, A.; Wei, G. J.; Bloomfield, V. A. J. Phys. Chem. 1983, 87, 3537.
31. Fendler, J. H. Acc. Chem. Res. 1980, 13, 7.
32. Okahata, Y.; Tanamachi, S.; Nagai, M.; Kunitake, Y. J. Colloid Interface Sci. 1981, 82, 401.
33. Franses, I.; Talmon, Y.; Scriven, L. E.; Davis, H. T.; Miller, W. G. J. Colloid Interface Sci. 1982, 86, 449.
34. Paradies, H. H. Angew. Chem. Int. Ed. Engl. 1982, 21, 765.

35. Kunitake, T.; Okahata, Y. Bull. Chem. Soc. Jpn. 1978, 51, 1877.
36. Fendler, J. H. In "Use of Ordered Media in Chemical Separations"; Hinze, W. L.; Armstrong, D. W., Eds.; American Chemical Society: Washington, D.C., 1987; Chapter 2.
37. Luisi, P. L.; Straub, B.E. "Reversed Micelles"; Plenum Press: New York, 1984.
38. Eicke, H. F. Pure Appl. Chem. 1981, 53, 1417.
39. Muller, N. J. Colloid Interface Sci. 1978, 63, 383.
40. Shinoda, K. "Solvent Properties of Surfactant Solutions"; Marcel Dekker, Inc.: New York, 1967.
41. Luisi, P. L.; Magid, L. J. CRC Crit. Rev. Biochem. 1986, 20, 409.
42. Hernandez-Torres, M. A.; Landy, J. S.; Dorsey, J. G. Anal. Chem. 1986, 58, 744.
43. Shinoda, K.; Friberg, S. "Emulsions & Solubilization"; Wiley: New York, 1986.
44. Shah, D. O. (Ed.) "Marco-and Microemulsions: Theory and Applications"; ACS Symposium Series 272; American Chemical Society: Washington, D.C., 1985.
45. Neustader, E. L. In "Surfactants"; Tadros, T. F., Ed.; Academic Press: New York, 1984, p. 277.
46. Holt, S. L. J. Dispersion Sci. & Tech. 1980, 1, 423.
47. Fendler, J. H. C & EN 1984, 62, 25-28; and references therein.
48. Menger, F. M.; Sato, K.; Martura, R. J. Phys. Chem. 1982, 86, 2463.
49. Hinze, W. L.; Singh, H. N.; Baba, Y.; Harvey, N. G. Trends Anal. Chem. 1984, 3, 193.
50. Berezin, I. V.; Martinek, K.; Yatsiminsky, A. K. Russ. Chem. Rev. 1973, 42, 778.
51. Hubig, S. M.; Dionne, B. C.; Rodgers, M. A. M. J. Phys. Chem. 1986, 90, 5873.
52. Costa, S. M. B.; Brookfield, R. L. J. Chem. Soc., Faraday Trans. 2, 1986, 82, 991.
53. Zultewicz, J. A.; Munoz, S. J. Phys. Chem. 1986, 90, 5820.
54. Miyaka, Y.; Shigeto, M.; Teramoto, M. J. Chem. Soc., Faraday Trans. 1, 1986, 82, 1515.
55. Ige, J.; Soriyan, O. J. Chem. Soc., Faraday Trans. 1, 1986, 82, 2011.
56. Otero, C.; Rodenas, E.; J. Phys. Chem. 1986, 90, 5771.
57. Blokhus, A. M.; Hoiland, H.; Backlund, S. J. Colloid Interface Sci. 1986, 114, 9.
58. Treiner, C.; Bocquet, J. F.; Pommier, C. J. Phys. Chem. 1986, 90, 3052.
59. Calvaruso, G.; Cavasino, F. P.; Di Dio, E. Inorg. Chim. Acta. 1986, 119, 29.
60. Gormally, J.; Sharma, S. J. Chem. Soc., Faraday Trans. 1, 1986, 82, 2497.

61. Cuccovia, I. M.; Schroter, E. H.; Monteiro, P. M.; Chaimovich,
     H. J. Org. Chem. 1978, 43, 2248.
62. Reisinger, M.; Lightner, D. A. J. Inclusion Phenom. 1985, 3,
     485.
63. Rico, I.; Lattes, A. J. Phys. Chem. 1986, 90, 5870.
64. Gilpin, R. K.; Kasturi, A.; Werner, G. J. Chromatogr. 1986,
     366, 293.
65. Manabe, M.; Tanizaki, Y.; Watanabe, H.; Niihama Kogyo Koto
     Semmon Gakko Kiyo, Rikogaku Hen 1983, 19, 50 [Chem. Abstr.,
     98:204875x].
66. Lissant, K. J.,; Ed. "Emulsions and Emulsion Technology",
     Part III, Marcel Dekker: New York, 1984.
67. Becker, P., Ed. "Encyclopedia of Emulsion Technology",
     Volumes 1 & 2, Marcel Dekker: New York, 1983; 1985.
68. Stranahan, J. L.; Deming, S. N. Anal. Chem. 1982, 54, 2251.
69. Tomlinson, E.; Jefferies, T. M.; Riley, C. M. J. Chromatogr.
     1978, 159, 315.
70. Fransson, B.; Wahlund, K. G.; Johansson, I. M.; Schill, G. J.
     Chromatogr. 1976, 125, 327.
71. Karger, B. L.; LePage, J. N.; Tanaka, N. In "High-Performance
     Liquid Chromatography"; Horvath, C., Ed.; Academic Press: New
     York, 1980; Vol. 1, 187.
72. Bidlingmeyer, B. A. LC Mag. 1983, 1, 344.
73. Burns, D. Anal. Proc. (London) 1982, 19, 355.
74. Knox, J. H.; Laird, G. R. J. Chromatogr. 1976, 122, 17.
75. Ghaemi, Y.; Wall, R. A. J. Chromatogr. 1980, 198, 397.
76. Wall, R. A. J. Chromatogr. 1980, 194, 353.
77. Bidlingmeyer, B. A.; Deming, S. N.; Price, W. P.; Sachok, W.
     P.; Petrusek, M. J. Chromatogr. 1979, 186, 419.
78. Horvath, C.; Melander, W.; Molnar, I.; Molnar, P. Anal. Chem.
     1977, 49, 2295.
79. Kraak, J. C.; Jonker, K. M.; Huber, J. F. K. J. Chromatogr.
     1977, 142, 671.
80. Knox, J. H.; Hartwick, R. J. Chromatogr. 1981, 204, 3.
81. Gloor, R.; Johnson, E. L. J. Chromatogr. Sci. 1977, 15, 413.
82. Sorel, R. H.; Hulshoff, A.; Snelleman, C. J. Chromatogr.
     Biomed. Appl. 1980, 221, 129.
83. Jira, T.; Beyrich, T.; Lemke, E. Pharmazie 1984, 39, 141.
84. Mullins, F. G. P., In "Use of Ordered Media in Chemical
     Separations"; Hinze, W. L.; Armstrong, D. W., Eds.; American
     Chemical Society: Washington, D.C., 1987, in press.
85. Girard, I.; Gonnet, C. J. Liq. Chromatogr. 1985, 8, 2035.
86. Armstrong, D. W.; Fendler, J. H. Biochim. Biophys. Acta 1977,
     478, 75.
87. Armstrong, D. W.; Terrill, R. Q. Anal. Chem. 1979, 51, 2160.
88. Armstrong, D. W.; McNeely, M. Anal. Lett. 1979, 12, 1285.
89. Armstrong, D. W.; Stine, G. Y. J. Amer. Chem. Soc. 1983, 105,
     2962.
90. Armstrong, D. W.; Bui, K. H. J. Liq. Chromatogr. 1982, 5,
     1043.

91. Armstrong, D. W.; J. Liq. Chromatogr. 1980, 3, 895.
92. Armstrong, D. W.; Spino, L. A.; Ondrias, M. R.; Findsen, E. W.
     J. Chromatogr. 1987, in press.
93. Berthod, A.; Armstrong, D. W. Anal. Chem. 1987, 59,
     submitted.
94. Armstrong, D. W.; Bui, K. H.; Barry, R. M. J. Chem. Educ.
     1984, 61, 457.
95. Armstrong, D. W.; Henry, S. J. J. Liq. Chromatogr. 1980, 3,
     657.
96. Armstrong, D. W.; Nome, F. Anal. Chem. 1981, 53, 1662.
97. Armstrong, D. W.; Hinze, W. L.; Bui, K. H.; Singh, H. N.
     Anal. Lett. 1981, 14, 1659.
98. Armstrong, D. W.; Stine, G. Y. J. Amer. Chem. Soc. 1983, 105,
     6220.
99. Armstrong, D. W.; Stine, G. Y. Anal. Chem. 1983, 55, 2317.
100. Armstrong, D. W.; Ward, T. J.; Berthod, A. Anal. Chem. 1986,
     58, 579.
101. Barth, H. G.; Barker, W. E.; Lochmuller, C. H.; Majors, R. E.;
     Regnier, F. E. Anal. Chem. 1986, 58, 228R.
102. Arunyanart, M.; Cline Love, L. J. Anal. Chem. 1984, 56, 1557.
103. Arunyanart, M.; Cline Love, L. J. Anal. Chem. 1985, 57, 2837.
104. Graglia, R.; Pramauro, E.; Pelizzetti, E. Ann. Chim. (Rome)
     1984, 74, 41.
105. Pramauro, E.; Pelizzetti, E. Anal. Chim. Acta 1983, 154, 153.
106. Borgerding, M. F.; Hinze. W. L. Anal. Chem. 1985, 57, 2183.
107. Mullins, F. G. P.; Kirkbright, G. F. Analyst 1984, 109, 1217.
108. Kirkbright, G. F.; Mullins, F. G. P. Analyst 1984, 109, 493.
109. Dorsey, J. G.; DeEchegaray, M. T.; Landy, J. S. Anal. Chem.
     1983, 55, 924.
110. Berthod, A.; Girard, I.; Gonnet, C. Anal. Chem. 1986, 58,
     1362.
111. Weinberger, R.; Yarmchuk, P.; Cline Love, L. J. Anal. Chem.
     1982, 54, 1552.
112. Hinze, W. L.; Fu, Z. C.; Sadek, F. unpublished results.
113. Yarmchuk, P.; Weinberger, R.; Hirsch, R. F.; Cline Love, L. J.
     J. Chromatogr. 1984, 283, 47.
114. Yarmchuk, P.; Weinberger, R.; Hirsch, R. F.; Cline Love, L. J.
     Anal. Chem. 1982, 54, 2233.
115. Terweij-Groen, C. P.; Heemstra, S.; Kraak, J. C. J.
     Chromatogr. 1978, 161, 69.
116. Vander Houwen, O. A.; Sorel, R. H.; Hulshoff, A.; Teeuwsen,
     J.; Indemenus, A. W.; J. Chromatogr. 1981, 209, 393.
117. Sorel, R. H.; Hulshoff, A.; Wiersema, S. J. Liq. Chromatogr.
     1976, 122, 17.
118. Knox, J. H.; Laird, G. R. J. Chromatogr. 1976, 122, 17.
119. Hansen, S. H.; Helboe, P.; Thomsen, M.; Lund, U. J.
     Chromatogr. 1981, 210, 453.
120. Hearn, M. T.; Grego, B. J. Chromatogr. 1984, 296, 309.
121. Hansen, S. H.; Helboe, P. J. Chromatogr. 1984, 285, 53.
122. Nowak, J. R. Z. Phys. Chem. (Leipzig) 1985, 266, 997.

123. Ottewill, R. H.; Rochester, C. H.; Smith, A. L. "Adsorption
      from Solution"; Academic Press: New York, 1983.
124. Gellan, A.; Rochester, C. H. J. Chem. Soc., Faraday Trans. 1
      1985, 81, 2235; 3109.
125. Harwell, J. H.; Hoskins, J. C.; Schechter, R. S.; Wade, W. H.
      Langmuir 1985, 1, 251.
126. Berthod, A.; Girard, I.; Gonnet, C. Anal. Chem. 1986, 58,
      1356.
127. Berthod, A.; Girard, I.; Gonnet, C. Anal. Chem. 1986, 58,
      1359.
128. Tramposch, W. G.; Weber, S. G. Anal. Chem. 1986, 58, 3006.
129. Hansen, S. H.; Helboe, P.; Lund, U. J. Chromatogr. 1983, 270,
      83.
130. El Rassi, Z.; Horvath, C. J. Chromatography 1985, 326, 79.
131. El Rassi, Z.; Horvath, C. Chromatographia 1982, 15, 75.
132. Roberts, B. L.; Scamehorn, J. F.; Harwell, J. H. In
      "Phenomena in Mixed Surfactant Systems"; Scamehorn, J. F.,
      Ed., ACS Symposium Series No. 311, American Chemical Society:
      New York, 1986, Ch. 15, p. 200.
133. Komarov, V. S.; Kuznetsova, T. F.; Barkatina, E. N. Vestsi
      Akad. Navuk BSSR, Ser. Khim. Navuk (Russ.) 1986, (#4), 22;
      [Chem. Abstr. 105: 140420y].
134. Hansen, S. H.; Helboe, P.; Thomsen, M. J. Chromatogr. 1986,
      360, 53.
135. Borgerding, M. F.; Hinze, W. L. "Abstracts of Papers - 191st
      ACS Meeting"; American Chemical Society: Washington, D.C.,
      1986, Abstr. No. 102.
136. Berthod, A.; Girard, I.; Gonnet, C. In "Use of Ordered Media
      in Chemical Separations", Hinze, W. L.; Armstrong, D. W. Eds.;
      American Chemical Society: Washington, D.C., 1987, in press.
137. Ghaemi, Y.; Wall, R. A. J. Chromatogr. 1981, 212, 272.
138. Djuric, Z.; Jovanovic, M. Ark. Farm. 1985, 35, 197; [Chem.
      Abstr. 104:155815j].
139. Landy, J. S.; Dorsey, J. G. Anal. Chim. Acta 1985, 178, 179.
140. Landy, J. S.; Dorsey, J. G. J. Chromatogr. Sci. 1984, 22, 68.
141. Dorsey, J. G.; Khaledi, M. G.; Landy, J. S.; Lin, J. L. J.
      Chromatogr. 1984, 316, 183.
142. Hernandez-Torres, M. A.; Landy, J. S.; Dorsey, J. G. Anal.
      Chem. 1986, 58, 744.
143. Dorsey, J. G. In "Use of Ordered Media in Chemical
      Separations"; Hinze, W. L., Armstrong, D. W., Eds.; American
      Chemical Society: Washington, D.C., 1987, in press.
144. DeLuccia, F. J.; Arunyanart, M.; Cline Love, L. J. Anal.
      Chem. 1985, 57, 1564.
145. Arunyanart, M.; Cline Love, L. J. J. Chromatogr. Biomed.
      Appl. 1985, 342, 293.
146. DeLuccia, F. J.; Arunyanart, M.; Yarmchuk, P.; Weinberger, R.;
      Cline Love, L. J. LC Mag. 1985, 3, 797.
147. Watson, I. D.; Stewart, M. J.; Farid, Y. Y. J. Pharm. Biomed.
      Anal. 1985, 3, 555.
148. Cline Love, L. J.; Zibas, S.; Noroski, J.; Arunyanart, M. J.
      Pharm. Biomed. Anal. 1985, 3, 511.

149. In "Research and Invention"; Research Corporation: Tucson, AZ, Fall 1986 issue, p. 2.
150. Rawat, J. P.; Singh, O. J. Indian Chem. Soc. 1986, 63, 248.
151. Liu, H.; Ling, J.; Yan, X. Zhejiang Gongxueyuan Xuebao 1985, 29, 39.
152. Barford, R. A.; Sliwinski, B. J. Anal. Chem. 1984, 56, 1554.
153. Graham, J. A.; Rogers, L. B. J. Chromatogr. Sci. 1980, 18, 614.
154. Qin, X.; Wu, C.; Chen, B. Acta. Pharmaceutica Sinica 1986, 21, 458.
155. Khaledi, M. G.; Dorsey, J. G. Anal. Chem. 1985, 57, 2190.
156. Kirkman, C. M.; Zu-Ben, C.; Uden, P. C.; Stratton, W. J. J. Chromatogr. 184, 317, 569.
157. Nagyvary, J.; Harvey, J. A.; Nome, F.; Armstrong, D. W.; Fendler, J. H. Precambrian Res. 1976, 3, 509.
158. Fendler, J. H.; Nome, F.; Nagyvary, J. J. Mol. Evol. 1975, 6, 215.
159. Armstrong, D. W.; Seguin, R.; Fendler, J. H. J. Mol. Evòl. 1977, 10, 241.
160. Singleton, J. A.; Pattee, H. E. J. Amer. Oil Chem. Soc. 1985, 62, 739.
161. Bedard, P. R.; Purdy, W. C. J. Liq. Chromatogr. 1986, 9, 1971.
162. Dydek, S. T.; Kehrer, J. P. LC Mag. 1984, 2, 536.
163. Igarashi, S.; Nakano, M.; Yotsuyanagi, T. Bunseki Kagaku 1983, 32, 68.
164. Farulla, E.; Iacobelli-turi, C.; Lederer, M.; Salvetti, F. J. Chromatogr. 1963, 12, 255.
165. Matson, R. S.; Goheen, S. C. LC·GC Mag. 1986, 4, 624.
166. Mabuchi, H.; Nakahashi, H. J. Chromatogr. 1981, 213, 275.
167. Simmonds, R. J.; Yon, R. J. Biochem. J. 1976, 157, 153.
168. Stahr, H. M.; Domoto, M. In "Planar Chromatography", Kaiser, R., Ed.; Springer-Verlag: New York, in press.
169. Mori, K.; "Abstracts of Papers - Eight International Symposium on Column Liquid Chromatography" 1984, Abstr. No. 4a-15, p. 23.
170. Regnier, F. Anal. Chem. 1983, 55, 1302A.
171. Carson, S. D.; Konigsberg, W. H. Anal. Biochem. 1981, 116, 398.
172. Castro, V.; Canselier, J. P. J. Chromatogr. 1986, 363, 139.
173. Wetlaufer, D. B.; Koenigbauer, M. R. J. Chromatogr. 1986, 359, 55.
174. Eicke, H. F.; Parfitt, G. D., Eds. "Interfacial Phenomena in Apolar Media"; Marcel Dekker: New York, 1986.
175. Lasovsky, J.; Grambal, F.; Petru, S. Acta Univ. Palacki. Olomuc, Fac. Rerum Nat. 1985, 82, 25; [Chem. Abstr. 104: 31118q].
176. Russo, M. V.; Goretti, G.; Liberti, A.; Laencina-Sanchez, J.; Flores, L. J. Essenze Deriv. Agrum. 1984, 54, 13.
177. Matsumaga, H.; Suzuki, T. M. Nippon Kagaku Kaishi 1986, (#7), 859.
178. Yeole, C. G.; Shinde, V. M. Indian J. Chem., Sect. A 1984, 23, 1053.
179. Naito, K.; Ogawa, H.; Moriguchi, S.; Takei, S. J. Chromatogr. 1984, 299, 73.

180. Wickramanayake, P. P.; Aue, W. A. Can. J. Chem. 1986, 64, 470.
181. Taguchi, K.; Hiratani, K.; Okahata, Y. J. Chem. Soc., Chem. Commun. 1986, 364.
182. Herries, D. G.; Bishop, W.; Richards, F. M. J. Phys. Chem. 1964, 68, 1842.
183. Martinek, K.; Yatsiminski, A. K.; Osipov, A. P.; Berezin, I. V. Tetrahedron 1973, 2, 963.
184. Zimina, T. M.; Maltsev, V. G.; Belenkii, B. G. J. High Resolut. Chromatogr. Chromatogr. Commun. 1986, 9, 111.
185. Booth, C.; Forget, J. L.; Lally, T. P.; Naylor, T. V.; Price, C. In "Chromatography of Synthetic and Biological Polymers: Column Packings, GPC, GF, and Gradient Elution", Epton, R., Ed., Ellis Horwood Ltd.: England, Volume I, Chapter 24, p. 324.
186. Spacek, P. J. Appl. Polym. Sci. 1986, 32, 4282.
187. Muccio, D. D.; DeLucas, L. J. J. Chromatogr. 1985, 326, 243.
188. Calam, D. H.; Davidson, J. J. Chromatogr. 1984, 296. 285.
189. Schreurs, V. V. A. M.; Boekholt, H. A.; Koopmanschap, R. E. J. Chromatogr. 1983, 254, 203.
190. Griffin, M. C. A.; Anderson, M. Biochim. Biophys. Acta 1983, 748, 453.
191. Budtov, V. P.; Domnicheva, N. A.; Podosenova, N. G. Kolloidn. Zh. 1984, 46, 614.
192. Goto, A.; Takemoto, M.; Endo, F. Bull. Chem. Soc. Jpn. 1985, 58, 247.
193. LePage, J. N.; Rocha, E. M. "Abstracts of Papers - Pittsburgh Conference", Abstr. No. 458, 1982.
194. Maley, F.; Guarino, D. U. Biochem. Biophys. Res. Commun. 1977, 77, 1425.
195. Dubin, P. L. In "Size Exclusion Chromatography"; Provder, T., Ed., ACS Symposium Series No. 138; American Chemical Society: Washington, D.C., 1980, Ch. 12, p. 225.
196. Terabe, S.; Otsuka, K.; Ichikawa, K.; Tsuchiya, A.; Andro, T. Anal. Chem. 1984, 56, 111.
197. Terabe, S.; Otsuka, K.; Ando, T. Anal. Chem. 1985, 57, 834.
198. Otsuka, K.; Terabe, S.; Ando, T. J. Chromatogr. 1985, 332, 219.
199. Otsuka, K.; Terabe, S.; Ando, T. J. Chromatogr. 1985, 348, 39.
200. Otsuka, K.; Terabe, S.; Ando, T. Nippon Kagaku Kaishi 1986, (#7), 950.
201. Burton, D. E.; Sepaniak, M. J.; Maskarinec, M. P. J. Chromatogr. Sci. 1986, 24, 347.
202. Burton, D. E.; Sepaniak, M. J.; Maskarinec, M. P. Chromatographia 1986, 21, 583.
203. Sepaniak, M. J.; Burton, D. E.; Maskarinec, M. P. In "Use of Ordered Media in Chemical Separations"; Hinze, W. L.; Armstrong, D. W., Eds., American Chemical Society: Washington, D.C., 1987, in press.
204. Sepaniak, M. J.; Cole, R. O. Anal. Chem. 1987, 59, 472.
205. Saitoh, T.; Hoshino, H.; Yotsuyanagi, T. Anal. Chem. 1987, 59, submitted.
206. Mullins, F. G. P.; Kirkbright, G. F. Analyst 1986, 111, 1273.

207. Spink, C. H.; Colgan, S. J. Phys. Chem. 1983, 87, 888.
208. Turro, N. J.; Somasundaran, P.; Waterman, K. C.; Chandar, P. "Abstracts of Papers"; Sixth International Symposium on Surfactants in Solution, August, 1986, Abstr. AA6R, p. 236.
209. Rao, I. V.; Ruckenstein, E. J. Colloid Interface Sci. 1986, 113, 375.
210. Kumar, J. L.; Mann, W. C.; Rozanski, A. J. Chromatogr. 1982, 249, 373.
211. Kunitake, T. In "Mod. Trends Colloid Sci.-Chem. Biol., Int. Symp. Colloid and Surf. Sci"; Eicke, H. F., Ed., Birkhaeuser: Switzerland, 1984, pp. 34-54.
212. Levitz, P.; Van Damme, H.; Keravis, D. Collect. Colloq. Semin. (Inst. Fr. Pet.) 1985, 42, 473; [Chem. Abstr. 105: 179033z].
213. Hitachi Ltd., Jpn. Kokai Tokkyo Koho, Patent #JP 59 10,849, 1984; [Chem. Abstr. 101: 35566v].
214. Terabe, S.; Utsumi, H.; Otsuka, K.; Ando, T.; Inomata, T.; Kuze, S.; Hanaoka, Y. HRC & CC 1986, 9, 666.
215. Shinoda, K.; Lindman, B. Langmuir 1987, 3, 135.
216. Zana, R. J. Chimie Phys. 1986, 83, 603.
217. Wu, J.; Harwell, J. H.; O'Rear, E. A. J. Phys. Chem. 1987, 91, 623.
218. Kim, Y, N.; Brown, P. R. J. Chromatogr. 1987, 384, 209.
219. Risby, T. H.; Jiang, L. Anal. Chem. 1987, 59, 200.
220. Ji, S. Fenxi Huaxue 1985, 13, 660; [Anal. Abstr., 6J16, p. 641].
221. Waterman, K. C.; Turro, N. J.; Chandor, P.; Somasundaran, P. J. Phys. Chem. 1986, 90, 6828; 1987, 91, 148.
222. Omori, H.; Amano, K.; Mori, K.; Yamazahi, S. J. High Resolut. Chromatogr. Chromatogr. Commun. 1987, 10, 47.
223. Deschamps, J. R. J. Liq. Chromatogr. 1986, 9, 1635.
224. Stratton, L. P.; Hynes, J. B.; Priest, D. G.; Doig, M. T.; Banon, D. A.; Asleson, G. L. J. Chromatogr. 1986, 357. 183.
225. McNeil, G. P.; Donnelly, W. J. J. Dairy Res. 1987, 54, 19.
226. Goto, A.; Nihei, M.; Endo, F. J. Phys. Chem. 1980, 84, 2268.
227. Teo, H. K.; Styring, M. G.; Yeates, S. G.; Price, C.; Booth, C. J. Colloid Interface Sci. 1986, 114, 416.
228. Tuong, T. D.; Hayano, S. Chem. Lett. (Japan) 1977, 1323.
229. Wilson, I. D. J. Chromatogr. 1986, 354, 99.
230. Grazhiuliene, S.; Nagy, V.; Orlova, T.; Kireiko, V. U.; Telegin, G. F. Mikrochim. Acta 1985, II, 153.
231. Yoshinaga, A.; Gohshi, Y. Bunseki Kagaku 1986, 35, 789.
232. Capacci, M. J.; Sepaniak, M. J. J. Liq. Chromatogr. 1986, 9, 3365.
233. Balchunas, A.; Capacci, M., Maskarinec, M.; Sepaniak, M. J. J. Chromatogr. Sci. 1985, 23, 381.
234. Altria, K. D.; Simpson, C. F. Anal. Proceedings 1986, 23, 453.
235. Cohen, A. S.; Terabe, S.; Smith, J. A.; Karger, B. L. Anal. Chem. 1987, 59, 1021.
236. Djuric, Z.; Jovanovic, M. Acta Pol. Pharm. 1986, 43, 52; [Chem. Abstr. 105: 232816k].
237. Suelter, C. H. "A Practical Guide to Enzymology"; John Wiley & Sons, Inc.: New York, 1985, p. 70.

72   ORDERED MEDIA IN CHEMICAL SEPARATIONS

238. Helenius, A.; Simons, K. Biochim. Biophys. Acta 1975, 29, 415.
239. Friberg, S. E., Ed. "Microemulsions: Structure and Dynamics";
     CRC Press: Florida, 1987; and references therein.
240. Gallacher, L. V. In "Solution Behavior of Surfactants"; Mittal,
     K. L.; Fendler, E. J., Eds.; Plenum Press: New York, 1982, p.
     791.
241. Osseo-Asare, K.; Keeney, M. E. Int. Solvent Extr. Conf., Proc.
     1980, 1, 80; and references therein.
242. Desando, M. A.; Walker, S.; Calderwood, C. H. J. Mol. Liq.
     1985, 31, 123.
243. Ivanov, I. M.; Zaitsev, V. P.; Batishcheva, E. K. Nauk. SSSR
     Seriya A, Khimi. Nauk 1980, (#1), 16.
244. Hinze, W. L.; Armstrong, D. W.; Singh, H. N. unpublished data.
245. Day, M. C. Pure Appl. Chem. 1977, 49, 75.
246. Product Bulletin, King Industries, Inc.: Norwalk, CT, 1978.
247. Hogfeldt, E.; Soldatov, V. S. J. Inorg. Nucl. Chem. 1979, 41,
     575.
248. Kuvaeva, Z. I.; Soldatov, V. S.; Hogfeldt, E. J. Inorg. Nucl.
     Chem. 1979, 41, 579.
249. Apanasenko, V. V.; Reznik, A. M.; Molochko, V. A.; Bukin, V. I.;
     Panich, R. M.; Golubkova, A. S. Izvestiya Akad. Nauk. SSSR,
     Seriya Khim. 1981, (#7), 1681.
250. Komarov, E. V.; Kopyrin, A. A.; Proyaev, V. V. Teoretich.
     Eksperimental. Khim. 1981, 17, 620.
251. Foune, P.; Bauer, D. C. R. Acad. Sci. Paris 1981, 292, 1077.
252. Bauer, D.; Foune, P.; Lemerle, J. C. R. Acad. Sci. Paris
     1981, 292, 1019.
253. Dalen, A.; Wijkstra, J.; Gerritsmu, K. W. J. Inorg. Nucl.
     Chem. 1978, 40, 875.
254. Ilic, M. Z.; Cattrall, R. W. J. Inorg. Nucl. Chem. 1981, 43,
     2855.
255. Inoune, K.; Nose, Y.; Watanabe, H. Bull. Chem. Soc. Jpn.
     1977, 50, 2793.
256. Keeney, M. E.; Osseo-Asare, K. Polyhedron 1982, 1, 453.
257. Kertes, A. S.; Grauer, F. J. Inorg. Nucl. Chem. 1978, 40,
     1781.
258. Good, M. L.; Srivastava, S. C. J. Inorg. Nucl. Chem. 1965,
     27, 2429.
259. Fredlund, F.; Hogfeldt, E.; Korshunova, T. A.; Soldatov, V. S.
     Chemica Scripta 1977, 11, 212.
260. Kuvaeva, Z. I.; Popov, A. V.; Soldatov, V. S.; Hogfeldt Solv.
     Extr. Ion Exch. 1986, 4, 361.
261. Morris, D. F. C.; Short, E. L. Electrochim. Acta 1984, 29,
     1083.
262. Apanasenko, V. V.; Reznik, A. M.; Bukin, V. I.; Panick, M.
     Zh. Neorg. Khimi. 1983, 28, 2895.
263. Product Bulletin (#S42-378), King Industries, Inc.: Norwalk,
     CT, 1978; and references therein.
264. Flett, D. S.; Melling, J.; Cox, M. In "Handbook of Solvent
     Extraction"; Lo, T. C.; Baird, M. H. I.; Hanson, C., Eds.;
     John Wiley & Sons, Inc.: New York, 1983, Ch. 24, p. 629.

265. Burns, D. T. Anal. Proc. 1982, 19, 355.
266. Dehmlow, E. F.; Makrandi, K. J. Chem. Res., Synop. 1986, (#1), 32.
267. Rakhman'ko, E. M.; Starobinets, G. L.; Polishchuk, S. V. Vestsi Akad. Navuk BSSR, Ser. Khim. Navuk 1986, (#1), 10.
268. Golovanov, V. I.; Golovanov, A. I.; Leukhin, S. G. Zh. Neorg. Khim SSSR 1986, 31, 2594.
269. Ivanov, I. M.; Zaitsev, V. P.; Batishcheva, E. K. Izv. Sib. Otd. Akad. Nauk SSSR, Ser. Khim. Nauk 1986, (#1), 16.
270. Nikolaeva, L. V.; Volkov, V. A. Kolloidn. Zh. 1983, 45, 590.
271. Tricaud, C.; Hipeaux, J. C.; Lemerle, J. Addit. Schmierst. Arbeitsfluessigkeiten, Int. Kolloq., 5th 1986, 2, 9/3/1 - 9/3/7; [Chem. Abstr. No. 105: 175533q].
272. Huettinger, K. J.; Schegk, J. R. VDI-Ber. 1981, 409, 489.
273. Pilipenko, A. T.; Zharinova, T. A.; Takubenko, L. N.; Falendysh, N. F. Dopov. Akad. Nauk Ukr. RSR, Ser. B: Geol., Khim. Biol. Nauki 1984, (#3), 48; [Chem. Abstr. No. 101: 178027f].
274. Biresaw, G.; Bunton, C. A. J. Phys. Chem. 1986, 90, 5854.
275. Fernandez, L. A.; Elizalde, M. P.; Castresana, J. M.; Aguilar, M.; Wingefors, S. Solvent Extr. Ion Exch. 1985, 3, 807.
276. Zheng, Y.; Osseo-Assare, K. Solvent Extr. Ion Exch. 1985, 3, 825.
277. Zaboshanskii, V. M.; Dobortvorskii, A. M.; Pinson, V. V.; Chulkov, V. P. Zh. Obshchei Khim. 1986, 56, 665.
278. Popov, S. O.; Bagreev, V. V. Zh. Neorg. Khim. 1986, 31, 635.
279. Golovanov, V. I.; Golvanov, A. I.; Leukhin, S. G. Zh. Neorg. Khim. 1986, 31, 2594.
280. Pattee, D.; Musikas, C.; Faure, A.; Chachafy, C. J. Less Common Met. 1986, 122, 295.
281. Muller, W.; Diamond, R. M. J. Phys. Chem. 1966, 70, 3469.
282. Hoh, Y. C.; Chuang, W. Y.; Wang, W. K. Hydrometallurgy 1986, 15, 381.
283. Cox, M.; Flett, D. S. In "Handbook of Solvent Extraction"; Lo, T. C.; Baird, M. H. I.; Hanson, C., Eds.; John Wiley & Sons: New York, 1983, Ch. 2.2, p. 53.
284. Jiang, Y.; Su, Y. Huagong Xuebao 1986, (#3), 328; [Chem. Abstr. 106: 48619t].
285. Fourre, P.; Bauer, D.; Lemerie, J. Anal. Chem. 1983, 55, 662.
286. Bhattacharyya, S. N.; Ganguly, K. M. Radiochimica Acta 1986, 39, 199.
287. Bhattacharyya, S. N.; Ganguly, K. M. Radiochimica Acta 1986 40, 17.
288. Dodov, V. B.; Trukhlyaev, P. S.; Kalinichenko, B. S.; Shovetsov, I. K. Radiokhimiya 1986, 28, 640.
289. Bhattacharyya, S.; Ganguly, B. J. Radioanal. Nucl. Chem. 1986, 98, 247.
290. Rajab, A.; Pareau, D.; Moulin, J. P.; Chesne, A. Bull. Soc. Chim. France 1987, (#1), 29.
291. Raieh, M. A.; El-Dessouky, M. M. J. Radioanal. Nucl. Chem. 1985, 96, 611.
292. Huang, T. C.; Juang, R. S. Ind. Eng. Chem. Fundam. 1986, 25, 752.

293. Yamada, H.; Hayashi, H.; Fujii, Y.; Mizuta, M. Bull. Chem.
      Soc. Japan 1986, 59, 789.
294. Yamada, H.; Adachi, K.; Fujii, Y.; Mizuta, M. Solvent Extr.
      Ion Exch. 1986, 4, 1109.
295. Kar, S. B.; Ray, U. S. J. Indian Chem. Soc. 1985, 62, 701.
296. Khadzhiev, D.; Kyuchukov, G.; Boyadzhiev, L. God. Vissh.
      Khim.-Tekhnol. Inst., Sofia 1984, 29, 394; [Chem. Abstr.
      103: 25446h].
297. Beneitez, P.; Ortiz, S. J.; Ortega, J. An. Quim., Ser. A.
      1985, 81, 65.
298. Gorelova, A. V.; Sukova, L. M.; Romantseva, T. I.; Kolenkova,
      M. A.; Chernova, O. P. Russ. J. Inorg. Chem. 1986, 31, 1189.
299. Yamashiji, Y.; Matsushita, T.; Shono, T. Polyhedron 1986,
      5, 1291.
300. Shmidt, V. S.; Shorohkov, N. A.; Nikitin, S. D. Zhurnal
      Neorgan. Khimii. 1986, 31, 998.
301. Ensor, D. D.; McDonall, G. R.; Pippin, C. G. Anal. Chem.
      1986, 58, 1814.
302. Ohashi, K.; Nakata, S.; Katsume, M.; Nakamura, K.; Yamamoto,
      K. Anal. Sci. 1985, 1, 467.
303. Ide, S.; Takagi, M. Anal. Sci. 1986, 2, 265.
304. Sato, T. Keikinzoku 1986, 36, 137; [Chem. Abstr. 106:23947k].
305. Reddy, M. L. P.; Rangamannai, B.; Reddy, A. S. Radioisotopes
      1985, 34, 675.
306. Vibhute, C. P.; Khopkar, S. M. Radioanal. Nucl. Chem. 1986,
      97, 3.
306. Ide, S.; Takagi, M. Anal. Sci. 1986, 2, 265.
307. Vibjute, C. P.; Khopkar, S. M. Bull. Chem. Soc. Jpn. 1986,
      59, 3229.
308. Shabama, R.; Khalifa, S. M.; Abdalla, S.; Aly, H. F. J.
      Radioanal. Nucl. Chem. 1986, 106, 55.
309. Torgov, V. G.; Tatarchuk, V. V.; Myul, P. Zh. Neorg. Khim.
      1987, 32, 126.
310. Bol'shova, T. A.; Ershova, N. I.; Kaplunova, A. M. Russ. J.
      Inorg. Chem. 1986, 31, 1035.
311. Nurtaeve, G. K.; Lobanov, F. I.; Zhurnal Neorgan. Khimmi
      1986, 31, 742.
312. Umetani, S.; Matsui, M.; Kawarro, H.; Nagai, T. Anal. Sci.
      1985, 1, 55.
313. Ivanova, S. N.; Sruzhimina, I. A.; Shuvaeva, O. V.;
      Yudelevich, I. G.; Yur'ev, G. S.; Kazakova, V. I. Russ. J.
      Inorg. Chem. 1986, 31, 1191.
314. Tsyganov, A. R.; Gurban, A. N. Russ. J. Inorg. Chem. 1986,
      31, 1091.
315. Caello, J.; Madariaga, J. M.; Muhammed, M.; Valiente, M.;
      Iturriaga, H. Polyhedron 1986, 11, 1845.
316. Alguacil, F. J.; Amer, S. Polyhedron 1986, 11, 1747.
317. Kimura, E.; Lin, Y.; Machida, R.; Zenda, H. J. Chem. Soc.,
      Chem. Commun. 1986, (#13), 1020.
318. Sakai, Y.; Nakamura, H.; Takagi, M.; Ueno, K. Bull. Chem.
      Soc. Japan 1986, 59, 381.
319. Yoshida, I.; Takashita, R. Anal. Sci. 1986, 2, 53.
320. Koshima, H.; Onishi, H. Nippon Kagaku Kaishi 1986, (#7),
      889.

321. Yoshida, I.; Hirasawa, J.; Tsumagari, H.; Ueno, K.; Takagi, M. Anal. Sci. 1986, 2, 447.
322. Puttemans, M.; Dryon, L.; Massart, D. L. Anal. Chim. Acta 1985, 15, 189.
323. Puttemans, M.; Dryon, L.; Massart, D. L. J. Pharm. Biomed. Anal. 1985, 3, 503.
324. Graef, V.; Banken, T.; Furuya, E.; Nishikaze, O. Fresenius' Z. Anal. Chem. 1986, 324, 289.
325. Kontturi, A. K.; Sundholm, G. Acta Chem. Scand., Ser. A. 1986, A40, 121.
326. Golovanov, V. I.; Ishimova, I. N. Zh. Neorg. Khim. 1987, 32, 141.
327. Skripchenko, A. S.; Soldatov, V. S. Dokl. Akad. Nauk BSSR 1987, 31, 59.
328. Mikucki, B. A.; Osseo-Asare, K. Hydrometallurgy 1986, 16, 209.
329. Bol'shakov, N. M.; Sinitsyn, N. M.; Buslaeva, T. M.; Samarova, L. V.; Ukraintseva, P. I. Dokl. Akad. Nauk SSSR 1986, 286, 926.
330. Wisniak, J.; Tamir, A., Eds. "Liquid-Liquid Equilibrium and Extraction" - A Literature Source Book, Part A; Elsevier Scientific Pub. Co.: New York, 1980; and references therein.
331. Wisniak, J.; Tamir, A., Eds. "Liquid-Liquid Equilibrium and Extraction" - Supplement 2; Elsevier Scientific Pub. Co.: New York, 1987; and references therein.
332. Li, N. N., Ed. "Recent Developments in Separation Science"; Volume VIII, CRC Press: Florida, 1986; and references therein.
332. Zolotov, Y. A.; Kuz'min, N. M.; Petrukhin, O. M.; Spivakov, B. Y. Anal. Chim. Acta 1986, 180, 137.
333. Osseo-Assare, K.; Keeney, M. E. Metall. Trans., B. 1980, 11B, 63.
334. Osseo-Asare, K.; Keeney, M. E. Sep. Sci. Technol. 1980, 15, 999.
335. Osseo-Asare, K.; Keeney, M. E. Solvent Extr. Ion Exchange 1983, 1, 337.
336. Samiolov, Y. M.; Yukhin, Y. M.; Shatskaya, S. S. Zh. Neorg. Khim. 1985, 30, 2053.
337. Keeney, M. E.; Osseo-Asare, K. Polyhedron 1982, 1, 541.
338. Apanasenko, U. V.; Rezik, A. M.; Bukin, V. I. Zh. Neorg. Khim. 1985, 30, 2187.
339. Osseo-Asare, K.; Renninger, D. R. Hydrometallurgy 1984, 13, 45.
340. Maksimovic, Z. B.; Puzic, R. G. J. Inorg. Nucl. Chem. 1972, 34, 1031.
341. Wang, S. M.; Chuang, H. C.; Tseng, C. L. J. Inorg. Nucl. Chem. 1975, 37, 1983.
342. Swarup, R.; Patil, S. K. J. Inorg. Nucl. Chem. 1976, 38, 1203.
343. Osseo-Asare, K.; Leaver, H. S.; Laferty, J. M. In "Process Fundam. Consid. Sel. Hydrometall. Syst.", Kuhn, L. M. C., Ed., Soc. Min. Eng. AIME: New York, 1981, p. 195; and references therein.
344. Tananaiko, M. M.; Philipenko, A. T. Zhur. Anal. Khim. 1977, 32, 430.

345. Diaz-Garcia, M. E.; Sanz-Medel, A. Talanta 1986, 33, 255.
346. Sato, T.; Nakamura, T. Bunseki 1986, (#1), 39.
347. Elizalde, M. P.; Castresana, J. M.; Aguilar, M.; Cox, M.
      Chemica Scripta 1985, 25, 300.
348. Ke-an, L.; Muralidharan, S.; Freiser, H. Solvent Ext. Ion
      Exch. 1985, 3, 895.
349. Nakache, E.; Dupeyrat, M.; Lemaire, J. J. Chim. Phys. France
      1986, 83, 339.
350. Takahashi, K.; Takeuchi, H. Kagaku Kogaku Ronbunshu 1985,
      11, 349.
351. Ajawin, L. A.; Demetriou, J.; Perez-Ortiz, E. S.; Sawistowski,
      H. Inst. Chem. Eng. Symp. Ser. 1984, 88, 183.
352. Yusupov, B. A.; Batrakova, L. K.; Kozlov, V. A. Kompleksn.
      Ispol'z Miner. Syr'ya 1986, (#5), 88; [Chem. Abstr., 105:
      67002g].
353. Urbanski, T. S. Isotopenpraxis 1986, 22, 288.
354. Bowers, C. B. Energy Res. Abstr. 1985, 10, Abstr. No. 12437;
      [Chem. Abstr. 103:44474g].
355. Con, T. H.; Anh, C. X. Tap Chi Hoa. Hoc 1984, 22, 7-9,32;
      [Chem. Abstr. 102: 188512v].
356. Lo, T. C.; Baird, M. H. I.; Hanson, C. "Handbook of Solvent
      Extraction"; John Wiley: New York, 1983; and references there-
      in.
357. Mitsubishi Metal Corp., Jpn. Kokai Tokkyo Koho JP (Patent)
      60 46,931 [85 46,931], 14 Mar 1985; [Chem. Abstr. 103:24443t].
358. Schulz, W. W.; Horwitz, E. P. J. Less-Common Met. 1986, 122,
      125.
359. Baba, Y.; Ohshima, M.; Inoune, K. Bull. Chem. Soc. Jpn. 1986,
      59, 3829.
360. Gaonker, A. G.; Neuman, R. D. Abstr. of Papers, Sixth Int'l
      Sym. on Surfactants in Solution, New Delhi, India, Aug., 1986,
      Abstr. No. MS19R, p. 19.
361. Ridgway, K.; Thorpe, E. E. In "Handbook of Solvent Extraction";
      Lo, T. C.; Baird, M. H. I.; Hanson, C., Eds., John Wiley & Sons:
      New York, 1983, Ch. 19, p. 583.
362. Pyatnitskii, I. V.; Savitskii, V. N.; Frankovskii, V. A.;
      Peleshenko, V. I.; Osadchii, V. I. Ukr. Khim. Zh. 1986, 52,
      44.
363. Janakiraman, B.; Sharma, M. M. Chem. Eng. Sci. 1982, 37,
      1497.
364. Li, H.; Chen, Z.; Zho, J. In "New Frontiers Rare Earth Sci.,
      Proc. Int. Conf. Rare Earth Dev. Appl", Xu, G.; Xiao, J., Eds.,
      Science Press: Beijing, China, 1985, 446.
365. Luisi, P. L.; Bonner, F. J.; Pellegrini, A.; Wiget, P.; Wolf,
      R. Helv. Chim. Acta 1979, 62, 740.
366. Goklen, K. E.; Hatton, T. A. Biotechnol. Prog. 1985, 1, 69.
367. Goklen, K. E.; Hatton, T. A. Abstr. Papers, 190th ACS Nat'l
      Meeting, Chicago, IL, 1985, Abstr. No. MBTD 128.
368. Goklen, K. E.; Hatton, T. A. Sep. Sci. Technol. 1987, in
      press.
369. Skew, E.; Goklen, K. E.; Hatton, T. A.; Chen, S. H. Biotech.
      Prog. 1986, 2, 175.
370. Leser, M. E.; Wei, G.; Luisi, P. L. Biochem. Biophys. Res.
      Commun. 1986, 135, 629.

371. Kadam, K. L. Enzyme & Microbial Techn. 1986, 8, 266.
372. Luisi, P. L.; Laane, C. Trends Biotechnol. 1986, 4, 153.
373. Martinek, K. Proc. FEBS Congr., 16th 1984, C, 211 [Chem.
     Abstr., 105: 931987].
374. Martinek, K.; Levashov, A.; Klyacho, N.; Khmel'nitskii, Y. L.;
     Berezin, I. V. Eur. J. Biochem. 1986, 155, 453.
375. Van't Riet, K.; Dekker, M. Eur. Congr. Biotechnol., 3rd 1984,
     3, 541.
376. Falbe, J.; Schmid, R. D. Fette, Seifen, Anstrichm. 1986, 88,
     203; [Chem. Abstr., 105: 77418m].
377. Hatton, T. A. In "Use of Ordered Media in Chemical Separ-
     ations"; Hinze, W. L.; Armstrong, D. W., Eds., American
     Chemical Society: Washington, D.C., 1987, in press.
378. Rahaman, R. S.; Cabral, J.; Hatton, T. A., manuscript in
     preparation.
379. Guiney, L.; Hart, J.; Hinze, W. L., unpublished data.
380. Laane, C.; Giovenco, S.; Dekker, M.; Baltussen, J.; van't Riet,
     K.; Bijsterbosch, B.; Veeger, C. Abstracts of Papers, Sixth
     Int'l. Symposium on Surfactants in Soln., New Delhi, India,
     August, 1986; Abstr. No. AA35L, p. 265.
381. Hatton, T. A. Paper at Conf. on Biodynamics, Bioreactors, &
     Downstream Processing, College of Engineering, Univ. Wisconsin,
     April, 1987.
382. Waks, M. Proteins: Struct., Funct., Genet. 1986, 1, 4;
     [Chem. Abstr., 105: 204819v].
383. Freeman, K. L.; Anderson, D. C.; Hughes, B.; Buffone, G. J.
     Clin. Chem. 1985, 31, 407.
384. Kyte, J. J. Biol. Chem. 1972, 247, 7642.
385. Jorgensen, P. L. Biochim. Biophys. Acta 1974, 356, 36.
386. Hall, J. D.; Crane, F. L. Biochim. Biophys. Acta 1972, 255,
     602.
387. Kuboyama, M.; Yong, F. K.; King, T. E. J. Biol. Chem. 1972,
     247, 6375.
388. Kagawa, Y. In "Methods in Membrane Biology", Korn, E. D., Ed.,
     Plenum Press: New York, 1974, Vol. 1, p. 201.
389. Kravtsova, V. V.; Kravtsov, A. V.; Yaroshenko, N. A.; Aryamova,
     Z. M. Neirokhimiya 1985, 4, 148.
390. Kirkpatrick, R.; Gordosky, S. E.; Marinetti, G. V. Biochim.
     Biophys. Acta 1974, 345, 154.
391. Helenius, A.; McCaslin, D. R.; Fries, E.; Tanford, C. Methods
     Enzymol. 1979, 56, 734.
392. Juliano, R. L. Curr. Top. Membr. Transp. 1978, 11, 107.
393. Holland, K.; Fu, Z. S.; Hinze, W. L., unpublished results.
394. Nagarajan, R.; Ruckenstein, E. Sep. Sci. Technol. 1981, 16,
     1429.
395. Friberg, S. E.; Mortensen, M.; Neogi, P. Sep. Sci. Technol.
     1985, 20, 285.
396. Neogi, P.; Kim, M.; Friberg, S. E. Sep. Sci. Technol. 1985,
     20, 613.
397. Plucinski, P. J. Heat Mass Transfer 1985, 28, 451.
398. Ward, A. J. I.; Carr, M. C.; Crudden, J. J. Colloid Interface
     Sci. 1985, 106, 558.
399. Lei, M. G.; Reeck, G. R. Cereal Chem. 1986, 63, 111 - 123.

400. Plucinski, P. K. Comun. Jorn. Com. Esp. Deterg. 1985, 16, 481.
401. Chaiko, M. A.; Nagarajan, R.; Ruckenstein, E. J. Colloid Interface Sci. 1984, 99, 168.
402. Helenius, A.; Simons, K. Biochim. Biophys. Acta 1975, 415, 29.
403. Batz, H. G. Topics in Biochem., Report from Boehringer Mannheim GmbH: Tutzing, W. Germany, pp. 2 - 11.
404. Venter, J. C.; Harrison, L. C. "Membranes, Detergents, and Receptor Solubilization", Alan R. Liss, Inc.: New York, 1984; and references therein.
405. Petrova, L. Y.; Glubokovskay, O. I.; Podsukhina, G. M.; Selezneva, A. A. Biokhimiya 1981, 46, 1570.
406. Giudicelli, J.; Bondouard, M.; Delque, P.; Vannier, C.; Sudaka, P. Biochim. Biophys. Acta 1985, 831, 59.
407. Guha, A.; Bach, R.; Konigsberg, W.; Nemerson, Y. Proc. Natl. Acad. Sci., USA 1986, 83, 299.
408. Negre, A. E.; Salvayre, R. S.; Dagan, A.; Gatt, S. Clin. Chim. Acta 1985, 149, 81.
409. Becker Freeman, K. L.; Anderson, D. C.; Hughes, B.; Buffone, G. J. Clin. Chem. 1985, 31, 407.
410. Yokoyama, S.; Ogawa, J. Jpn. Kokai Tokkyo Koho JP 43,986, 1986; [Chem. Abstr. 104: 205557z].
411. Kravtsova, V. V.; Kratsov, A. V.; Yaroshenko, N. A.; Aryamova, Z. M. Neirkhimiya 1985, 4, 148; [Chem. Abstr. 103: 174907m].
412. Trokhimenko, E. P.; Keisevich, L. V.; Pen'kovskaya, N. P.; Lozinskii, M. O.; Sereda, A. G. Otkrytiya Izobret., Prom. Obraztsy, Tovarnye Znaki 1984, (# 11), 82; [Chem. Abstr. 101: 35571t].
413. Green, T. R.; Wu, D. E. Biochim. Biophys. Acta 1985, 831, 74.
414. Hempelmann, E.; Putfarken, B.; Rangacheri, K.; Wilson, R. J. M. Parasitology 1986, 92, 305.
415. Miller, M. J.; Gennis, R. B. Methods Enzymol. 1986, 126, 87.
416. Matsushita, K.; Patel, L.; Kaback, H. R. Methods Enzymol. 1986, 126, 113.
417. Gardnier, M. A.; Peace, D. M. J. Assoc. Off. Anal. Chem. 1986, 69, 712.
418. Laughlin, R. G. In "Advances in Liquid Crystals", Brown, G. H., Ed., Academic Press: New York, 1978, Vol. 3, pages 41, 76, 103.
419. DeGiorgio, V.; Piazza, R.; Corti, M.; Minero, C. J. Chem. Phys. 1985, 82, 1025.
420. Corti, M.; Minero, C.; DeGiorio, V. J. Phys. Chem. 88, 309.
421. Corti, M.; DeGiorio, V.; Hayter, J. B.; Zulauf, M. Chem. Phys. Lett. 1984, 109, 580.
422. Goldstein, R. E. J. Chem. Phys. 1986, 84, 3367.
423. Zana, R.; Weill, C. J. Physique Lett. 1985, 46, L-953.
424. Kjellander, R. J. Chem. Soc., Faraday Trans. 2 1982, 78, 2025.
425. DeGiorgio, V. In "Physics of Amphiphiles: Micelles, Vesicles, and Microemulsions"; DeGiorgio, V.; Corti, Eds. North-Holland: Amsterdam, 1985, p. 303; and references therein.

426. Di Meglio, J. M.; Paz, L.; Dvdaitzky, M.; Taupin, C. J. Phys. Chem. 1984, 88, 6036.
427. Schott, H.; Han, S. K. J. Pharm. Sci. 1975, 64, 658.
428. Nilsson, P. G.; Wennerstrom, H.; Lindman, B. Chemica Scripta 1983, 25, 67.
429. Watanabe, H. In "Solution Behavior of Surfactants", Vol. 2; Mittal, K. L.; Fendler, E. J., Eds.; Plenum Press: New York, 1982, 1305.
430. Kawamorita, S.; Watanabe, H.; Haraguchi, K. Anal. Sci. 1985, 1, 41.
431. Watanabe, H.; Tanaka, H. Talanta 1978, 25, 585.
432. Ishii, H.; Miura, J.; Watanabe, H. Bunseki Kagaku 1977, 26 252.
433. Watanabe, H.; Yamaguchi, N.; Tanaka, H. Bunseki Kagaku 1979, 28, 366.
434. Kawamorita, S.; Watanabe, H.; Haraguchi, K. Proc. Symp. Solv. Extr. 1984, 75-80.
435. Hoshino, H.; Saitoh, T.; Taketomi, H.; Yotsuyanagi, T.; Watanabe, H.; Tachikawa, K. Anal. Chim. Acta 1983, 147, 339.
436. Qi, W. B.; Fu, K. Fenxi Huaxue 1983, 147, 339.
437. Pramauro, E.; Minero, C.; Pelizzetti, E. In "Use of Use of Ordered Media in Chemical Separations", Hinze, W. L.; Armstrong, D. W., Eds.; American Chemical Society: Washington, D. C., 1987, in press.
438. Takahasho, H.; Kuwamura, T. Bull. Chem. Soc. Jpn. 1973, 46, 623.
439. Bordier, C. J. Biol. Chem. 1981, 25, 1604.
440. Kippenberger, D. J.; Morris, M. D.; Fu, Z. S.; Singh, H. N.; Hinze, W. L. Abstr. of Papers, 191st National American Chemical Society Meeting, New York, 1986, Abstr. No. ANAL 117.
441. Koningsveld, R.; Kleintjens, L. A. British Polym. J. 1977, 212.
442. Schick, M. J., Ed. "Nonionic Surfactants: Physical Chemistry" Marcel Dekker, Inc.: New York, 1987; and references therein.
443. Koshima, H.; Onishi, H. Nippon Kagaku Kaishi 1986, (#7), 889; [Chem. Abstr., 105: 100785q].
444. Heusch, R.; Bayer, A. G. Abstracts of Papers, 6th Int'l Symp. on Surfactants in Soln., New Delhi, India, Aug., 1986; Abstr. No. AA14L, p. 244.
445. Vassiliades, A. E. In "Cationic Surfactants", Jungermann, E., Ed.; Marcel Dekker, Inc.: New York, 1970, Chapter 12, p. 387.
446. Tanake, K. Hyomen 1980, 18, 556.
447. Libackyj, A. In "Ordering Two Dimensions, Proc. Int'l. Conf." Sinha, S. K., Ed., Elsevier: New York, 1980, p. 459.
448. Sineva, A. V.; Markina, Z. N. Kolloidn. Zh. 1979, 41, 283.
449. Lemordant, D.; Gaborland, R. C. R. Acad. Sc. Paris 1981, 296, 981.
450. Siano, D. B.; Bock, J. J. Colloid Interface Sci. 1982, 90, 359.
451. De Trobriand, A. M. INIS Atomindex 1980, 11, No. 521209; Report CEA-R-5009, 18pp, 1979.
452. Charewicz, W. A.; Strzelbicki, J. J. Chem. Technol. Biotechnol. 1979, 29, 149.

453. Dubin, P. L.; Ross, T. D.; Sharma, I.; Yegerlehner, B. E. In
      "Use of Ordered Media in Chemical Separations", Hinze, W. L.;
      Armstrong, D. W., Eds.; American Chemical Society: Washington,
      1987, in press.
454. Dunn, R. O.; Scamehorn, J. F.; Christian, S. D. Sep. Sci.
      Technol. 1985, 20, 257.
455. Smith, G. A.; Christian, S. D.; Tucker, E. E.; Scamehorn, J.
      F. In "Use of Ordered Media in Chemical Separations", Hinze,
      W. L.; Armstrong, D. W.; Eds.; American Chemical Society:
      Washington, D.C., 1987, in press.
456. Scamehorn, J. F.; Ellington, R. T.; Christian, S. D.; Penney,
      B. W.; Dunne, R. O.; Bhat, S. N. AIChE Symp. Ser. 1986, 82,
      48; [Chem. Abstr., 106: 72330n].
457. Gibbs, L. L.; Scamehorn, J. F.; Christian, S. D. J. Membrane
      Sci. 1987, 30, 67.
458. Smith, G. A.; Christian, S. D.; Tucker, E. E.; Scamehorn, J. F.
      J. Soln. Chem. 1986, 15, 519.
459. Christian, S. D.; Smith, G. A.; Tucker, E. E.; Scamehorn, J. F.
      Langmuir 1985, 1, 564.
460. Sata, T.; Hanada, F.; Mizutani, Y. J. Membr. Sci. 1986, 28,
      151.
461. Li, N. N. U.S. Patent, No. 3,410,794, Nov. 12, 1968; AIChEJ
      1971, 17, 459.
462. Strzelbicki, J.; Charewicz, W. J. Inorg. Nucl. Chem. 1978,
      40, 1415.
463. Akiba, K.; Hashimoto, H. Anal. Sci. 1986, 2, 541.
464. Loiacono, O.; Drioli, E.; Molinari, R. J. Membr. Sci. 1986,
      28, 123.
465. Macasek, F.; Rajec, P.; Rehacek, V.; Vu Ngoc Anh, T.;
      Popovnakova, T. J. Radioanal. Nucl. Chem. 1985, 96, 529.
466. Guedj, P. Doc. B.R.G.M. 1983, 79, 73; [Chem. Abstr., 103:
      40350r].
467. Fujinawa, K.; Morishita, T.; Hozawa, M.; Ino, H. Kagaku
      Kogaku Ronbunshu 1985, 11, 293.
468. Dickens, N.; Rose, M.; Davies, G. A. Inst. Chem. Eng. Symp.
      Ser. 1984, 88, 69.
469. Danesi, P. R.; Chiarizia, R.; Rickert, P.; Horwitz, E. P.
      Solvent Extr. Ion Exch. 1985, 3, 111.
470. Plucinski, P. J. Membr. Sci. 1985, 23, 105.
471. Zhang, R.; Wang, D. Mo Fenli Kexue Yu Jishu 1985, 5, 70;
      [Chem. Abstr., 105: 9713d].
472. Lieberwirth, I.; Moehle, L.; Pfestorf, R. Wiss. Z. Karl-Marx
      Univ. Leipzig, Math.-Naturwiss. Reihe 1986, 35, 451; [Chem.
      Abstr., 106: 23587t].
473. Armstrong, D. W.; Li, W. Abstracts of Papers, 1987 Pittsburgh
      Conference, Atlantic City, NJ, March, 1987, Abstr. No. 539.
474. Krovvidi, K. R.; Stroeve, P. J. Colloid Interface Sci. 1986,
      110, 437.
475. Ho, W. S.; Li, N. N. NATO Conf. Ser. 6 1984, 6, 555 - 597.
      [Chem. Abstr. 103: 39115m].
476. Morf, W. W.; Huser, M.; Lindemann, B.; Schulthess, P.; Simon,
      W. Helv. Chim. Acta 1986, 69, 1333.
477. Nishiki, T.; Kataoka, T. Bull. Univ. Osaka Prefect., Ser A
      1985, 34, 97; [Chem. Abstr. 105: 66999a].

478. Gu, Z.; Zhang, H.; Wasan, D. T.; Li, N. N. Huagong Xuebao 1986, (#1), 1; [Chem. Abstr., 105: 8543m].
479. Yagodin, G. A.; Ivankhno, S. Y.; Gusev, Y.; Levkin, A. V. Dokl. Akad. Nauk. SSSR 1982, 266, 1198.
480. Li, N. N.; Navratil, J. D., Eds. "Recent Developments in Separation Science", CRC Press: Florida, Volume. IX, 1986; and references therein.
481. Starzak, M. E. "The Physical Chemistry of Membranes", Academic Press: New York, 1984; and references therein.
482. Armstrong, D. W.; Hinze, W. L.; Bui, K.; Singh, H. N. Anal. Lett. 1981, 14, 1659.
483. Alak, A.; Heilweil, E.; Hinze, W. L.; Oh, H.; Armstrong, D. W. J. Liq. Chromatogr. 1984, 7, 1273.
484. Cline Love, L. J.; Skrilec, M.; Habarta, J. G. Anal. Chem. 1980, 52, 754.
485. Weinberger, R.; Yarmchuk, P.; Cline Love, L. J. Anal. Chem. 1982, 54, 1552.
486. Nelson, P. E. J. Chromatogr. 1984, 298, 59.
487. Cline Love, L. J.; Weinberger, R.; Yarmchuk, P. In "Surfactants in Solution", Vol. 2; Mittal, K. L.; Lindman, B., Eds. Plenum Press: New York, 1984, 1139.
488. Khaledi, M. G.; Dorsey, J. G. Anal. Chem. 1985, 57, 2190.
489. Schwartz, D. P.; Sherman, J. T. J. Chromatogr. 1982, 240, 206.
490. Kirkman, C. M.; Zu-Ben, C.; Uden, P. C.; Stratton, W. J.; Henderson, D. E. J. Chromatogr. 1984, 317, 569.
491. Bedard, P. R.; Purdy, W. C. J. Liq. Chromatogr. 1986, 9, 1971.
492. Armstrong, D. W.; Spino, L. A.; Ondrias, M. R.; Findsen, E. W. J. Chromatogr. 1986, 369, 227.
493. Hinze, W. L.; Fu, Z.; Reihl, T. E., unpublished results.
494. Armstrong, D. W.; Spino, L. A.; Ondrias, M. R.; Findsen, E. W. J. Amer. Chem. Soc. 1986, 108, 5646.
495. Armstrong, D. W.; Spino, L. A.; Vo-Dinh, T.; Alak, A. Spectroscopy 1987, 2, 54.
496. Yuan, Y. Fenxi Huaxue 1986, 14, 426.
497. De Lima, C. G.; Andino, M. M.; Winefordner, J. D. Anal. Chem. 1986, 58, 2867.
498. Yuan, Y. Fenxi Huaxue 1986, 14, 426.
499. Endo, K.; Igarashi, S.; Yotsuyanagi, T. Chem. Lett. Jpn. 1986, (#10), 1711.
500. Napper, D. N. "Polymeric Stabilization of Colloidal Dispersions", Academic Press: New York, 1984.
501. Neustadter, E. L. In "Surfactants", Tadros, T. F., Ed., Academic Press: New York, 1984, p. 277.
502. Harwell, J. H.; Helfferich, F. G.; Schechter, R. S. AIChE J. 1982, 28, 448.
503. Rosano, H. L.; Clausse, M., Eds. "Microemulsion Systems", Marcel Dekker, Inc.: New York, 1987.
504. Nomura, T.; Umehara, T.; Takahashi, Y.; Yamamoto, M. Jpn. Kokai Tokkyo Koho JP, No. 61 29,613, 10 Feb. 1986; [Chem. Abstr., 105: 29194p].
505. Hjelmeland, L. M.; Chrambach, A. Electrophoresis 1981, 2, 1.

506. Cross, J., Ed. "Anionic Surfactants", Marcel Dekker, Inc.: New York, 1977.

507. Cross, J., Ed. "Nonionic Surfactants", Marcel Dekker, Inc.: New York, 1986.

508. Lunkenheimer, K.; Miller, R.; Kretzschmar, G.; Lerche, K. H.; Becht, J. Colloid Polym. Sci. 1984, 262, 662.

509. Titkov, S. N.; Kozlov, S. S.; Shevchenko, E. F.; Panteleeva, N. N.; Solov'ev, E. I.; Tolkachev, M. D.; Denkevich, T. E.; Nevel'son, I. S. U.S.S.R. SU 1,214,594, 1986; [Chem. Abstr., 105: 11516k].

510. "Previews", Pierce Chem. Co.: Rockford, IL, 1985, p. 6.

511. Elizinga, M.; Phelan, J. J. Proc. Natl. Acad. Sci. 1984, 81, 6599.

512. Laursen, S. E.; Knull, H. R.; Belknap, J. K. Anal. Biochem. 1986, 153, 387.

513. Holloway, P. W. Anal. Biochem. 1973, 53, 304.

514. Kapp, O. H.; Vinogradov, S. N. Anal. Biochem. 1978, 91, 230.

515. Jaeger, D. A.; Robertson, R. E. J. Org. Chem. 1977, 42, 3298.

516. Jaeger, D. A.; Ward, M. D. J. Org. Chem. 1982, 47, 2221.

517. Jaeger, D. A.; Frey, M. R. J. Org. Chem. 1982, 47, 311.

518. Jaeger, D. A.; Finley, C. T.; Walter, M. R.; Martin, C. A. J. Org. Chem. 1986, 51, 3956.

519. Takeda, T. Kagaku to Kogyo (Osaka) 1986, 60, 48; [Chem. Abstr., 105: 148995)

520. Handtke, K. D.; Stanke, R.; Keller, J. Ger. Offen. DE 3,332,588; 31 Mar 1983; [Chem. Abstr. 98: 177849h].

521. Ryabchenko, N. P.; Konstantinov, E. N.; Lyubchenkov, P. P.; Sukhina, M. I.; Belokhvostikov, V. I. U.S.S.R. SU 828,463, 23 Sept. 1985; [Chem. Abstr. 104: 52526m].

522. Shimshick, E. J. ChemTech 1983, 13, 374.

523. Brady, J. E. In "Research & Invention Newsletter - List of Grants", Research Corporation: Tucson, AZ, (Spring, 1986), p. 4.

524. Smith, R. D.; Yonker, C. R. (from Pacific Northwest Labs) In "Summaries of FY 1986 Research in Chemical Sciences" Report, US Department of Energy: Washington, D.C., Sept. 1986, p. 28.

525. Olesik, S. V.; Kim, H. S. "Titles from Preliminary Program" 1986 Pittsburgh Conference, March, 1986, Title No. 1048, p. 30.

526. Brady, J. E. In "Chemistry Department Brochure - Univ. Pittsburgh", 1987.

527. Berthod, A.; Armstrong, D. W. Anal. Chem. 1987, submitted.

RECEIVED April 16, 1987

Chapter 2

Membrane Mimetic Separations

Janos H. Fendler

Department of Chemistry, Syracuse University, Syracuse, NY 13244-1200

Development of new separation techniques requires a
fundamental understanding of the relationship between
molecular structures and permeabilities. Initiation
of interdisciplinary researches in biology, biophysics, polymer and colloid chemistry is proposed to
provide the insight to membrane transport processes
at the molecular level. Mother nature's most
talented transporter - the biological membrane -
should inspire this endeavor. Following a survey of
the properties of, and recognized transport mechanisms in, biomembranes, membrane mimetic chemistry
is introduced to serve as a bridge between biological
and polymeric membranes. Surfactant aggregates -
micelles, monolayers, organized multilayers
(Langmuir-Blodgett films), bilayer lipid membranes
(BLMs), vesicles and polymerized vesicles - are
shown to be the media in membrane mimetic chemistry.
Properties of these organized surfactant assemblies
are summarized. Emphasis is placed on our current
research on the potential use of BLMs to reconstitute
active and transport systems and on the development
of their simultaneous electrical and spectroscopic
measurements.

Micelles and other organized surfactant aggregates are increasingly
utilized in analytical applications (1). They interact with
reagents and alter spectroscopic and electrochemical properties
which, in turn, often results in increased sensitivities. Organized
assemblies have also been employed in separation processes. Gas,
liquid and thin layer micellar chromatographic techniques have been
developed (2).
    Realizing the full potential of organized assembly mediated
separations necessitates, I believe, well conceived and well
executed interdisciplinary researches. The purpose of this
presentation is to stimulate such interdisciplinary approaches. Our

0097-6156/87/0342-0083$06.25/0
© 1987 American Chemical Society

starting point will be mother nature's most talented transporter -
the biological membrane. Following a brief description of the
biological membrane (in the section on Biological Membranes), the
recognized transport mechanisms will be delineated therein (section
on Recognized Transport Mechanisms Across Biological Membranes),
The section on Membrane Mimetic Chemistry will discuss the
philosophy of the membrane mimetic approach and the most frequently
used mimetic systems. The section on Simultaneous Electrical and
Spectroscopic Measurements of BLMs will highlight our current
researches on BLM spectroscopy. The treatments will be, of course,
illustrative rather than comprehensive.

Biological Membranes

Biological membranes define the very existence of cells. They
provide compartments for the different components of the living
system; interact with, transport and are permeable to substrates.
They are involved in lipid and protein syntheses, energy transduction, ion and group transport, information transmission and
molecular and cellular recognition. These multitude of activities
are accomplished by the unique morphology of the biological
membrane and by its ability to affect the transport of species by
different mechanisms.
     Cell membranes are composed of 25-75% lipids, 25-75% proteins
and less than 10% carbohydrates. The organization of these
components in the membrane is best described in terms of the
bilayer-lipid - globular-protein "fluid mosaic" model (3,4). As
illustrated in Figure 1, the lipids (phospholipids and/or glyco-
lipids) are arranged in bilayers with their polar headgroups
exposed to the exterior surface of the membrane. Proteins are
either a peripheral or integral part of the membrane. The former,
attached electrostatically, is easily dissociated from the membrane
by changing the pH or the ionic strength of the solution. Integral
proteins partially intercalate the membrane or fully span the
bilayer (transmembrane protein). Globular proteins are partially
embedded into one or the other side of the membrane and form a
mosaic pattern with the lipid headgroups. The depth of incorporation depends upon the size of the globular protein, its hydrophobicity and charge distribution. An important requirement of the
fluid mosaic model is the dynamic nature of the lipid-protein
interactions in the membrane. Proteins may rotate around their
axes, diffuse laterally in the plane of the membrane or move across
the bilayer. Additionally, they may undergo vibrational and
conformational changes. Being less than categorical in describing
protein mobilities has been intentional. While most proteins move
about, some cannot freely diffuse in the membrane under physiological conditions.
     The lipids themselves are highly mobile. Steady state and
time resolved spectroscopy (absorption, emission, ir, raman, nmr,
epr) and anisotropy measurements have revealed rotational,
vibration and segmental motions of the headgroups and the hydrocarbon tails of the lipids. Translocation of a lipid from one half
of the bilayer to the other, ("flip-flop") as well as intermembrane

(or intervesicular) lipid exchanges have also been recognized. Figure 2 illustrates some of the motions of lipids.

Proteins and lipids interact cooperatively in the membrane. The type(s) and state(s) of lipids influence the mobility and conformation of the proteins in the membrane matrix. This, in turn, may well alter the properties of the membrane proteins. Similarly, proteins affect the phase behavior of the lipids and/or promote domain formation in membranes containing mixtures of lipids. Morphological alteration of the lipid architecture leads to changes in the membrane permeability.

Phase transition is an important property of membranes. Below the phase transition temperature, lipids are tilted and highly ordered. They are in their solid or "gel" state. Increasing the temperature leads to a pre-transition, characterized by periodic undulations and straightening of the hydrocarbon chain. Further increase of the temperature causes the main phase transition. Above the main phase transition temperature, lipids are fluid or "liquid crystalline." Figure 3 shows the phase diagram for the interaction of water with a lipid as well as its inferred arrangements in a model membrane (5). Phase transitions in membranes and membrane models have been extensively studied by spectroscopic techniques and by differential scanning calorimetry.

Most membranes are osmotically active. They shrink if electrolytes are added externally. They swell if placed in a solution which is more dilute than their internal electrolyte concentrations.

Most membranes are asymmetric with respect to the distribution of lipids, charges and proteins between their exteriors and interiors. Uneven distribution of ions between the outside and the inside of membranes is responsible, at least in part, for membrane potentials. The inside of living cells (cytoplasm, for example) is typically more negative than the extracellular medium. This difference in charges is referred to as the resting or membrane potential. Transient changes in the membrane potential, caused by reversible charge redistributions, are responsible for information and impulse transmission in nerve and muscle fibers. There is another important asymmetry in membranes: the segregation of certain lipids (phase separation) giving rise to domains. The precise function of domains has not been elucidated.

Emphasis is placed here on features of the biological membranes which are implicated in substrate transport. The lipid bilayer in the "gel" state, in the absence of additives, forms an effective barrier against polar ions and water soluble substrates. Changing the fluidity, by phase transition (induced by temperature changes and/or by the addition of foreign ions or molecules) or by the incorporation of additives (cholesterol, for example), profoundly influences the structure and, hence, the transport properties of membranes. This, and the presence of channel or pore forming peptides or proteins, opens the door to a number of transport mechanisms which will be summarized in the following section.

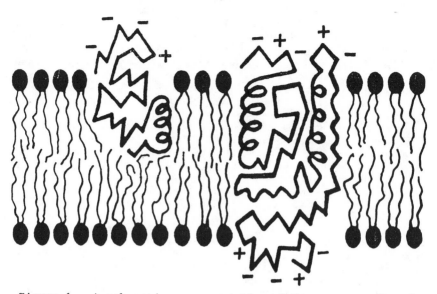

Figure 1. A schematic representation of the cross section of
the lipid-globular protein mosaic model of membrane structure.
The globular proteins (with dark lines denoting the polypeptide
chain) are amphipathic molecules with their ionic and highly
polar groups exposed at the exterior surfaces of the membranes;
the degree to which these molecules are embedded in the membrane
is under thermodynamic control. The bulk of the phospholipids
(with filled circles representing their polar head groups and
thin wavy lines their fatty acid chains) is organized as a
discontinuous bilayer.

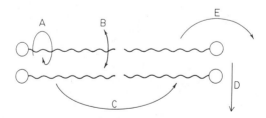

Figure 2. An oversimplified representation of molecular motions
in liposome bilayers. Individual lipids can rotate (A), undergo
sequential motion (B), flip-flop (C), undergo lateral diffusion
(D), or intervesicle exchange (E).

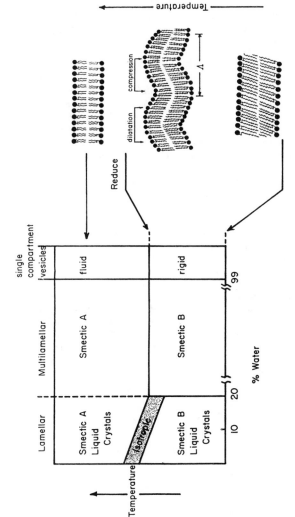

Figure 3. Schematic representation of a phospholipid-water phase diagram. The temperature scale is arbitrary and varies from lipid to lipid. For the sake of clarity phase separations and other complexities in the 20-99% water region are not indicated. Structures proposed for the phospholipid bilayers at different temperatures are shown on the right-hand side. At low temperature, the lipids are arranged in tilted one-dimensional lattices. At the pre-transition temperature, two-dimensional arrangements are formed with periodic undulations. Above the main phase, transitions lipids revert to one-dimensional lattice arrangements, separated somewhat from each other, and assume mobile liquid-like conformations.

Recognized Transport Mechanisms Across Biological Membranes

Transport across biological membranes is classified according to the thermodynamics of the process. Passive transport is a thermodynamically downhill process; the species move toward the equilibrium. The driving force for the passive transport is the potential difference between the two sides of the membrane. Active transport is a thermodynamically uphill process, it is coupled to a chemical reaction and is driven by it. The following transport mechanisms have been recognized:

Passive Transport. Transport by simple diffusion: This mode of transport is available for apolar molecules. Permeation is predominantly governed by partitioning of the substrate between the lipid and water. The membrane simply acts as a permeability barrier; small molecules pass more easily than large ones. The transport is explained in terms of a simple diffusion model involving three steps: passage of the substrate from the exterior into the membrane, diffusion through the membrane, and passage out of the membrane.

Transport by facilitated diffusion: A large number of molecules and ions were shown to permeate membranes considerably faster than expected from their lipid-water partitioning behavior. This led to the recognition of additional transport mechanisms. Systematic investigations of permeability rates in membranes, reconstituted membranes, and membrane models as functions of the temperature; of the nature and concentration of the permeant; in the absence and in the presence of additives, suggested three different facilitated passive transport mechanisms:

1) Carrier mediated transport - substrates are transported across the membrane by a diffusable carrier, typically an enzyme. Once again, there are three steps: complexation of the substrate with the carrier on one-side of the membrane, diffusion of the substrate-carrier complex to the other side and decomplexation:

$$
\begin{array}{ccccc}
S + & ES & \rightleftharpoons & ES & + S \\
 & \updownarrow & & \updownarrow & \\
 & E & \rightleftharpoons & E & \\
\end{array}
\qquad (1)
$$

The sugar-transport system is the most often cited example for the carrier mediated facilitated transport of a covalent molecule. Transport of sugars into the red blood cells is passive (it occurs only in the presence of a concentration gradient), selective (D-glucose is transported, while L-glucose is not), and the kinetics show a saturation behavior (observed typically for enzyme mediated interactions). These observations are in support of a facilitated passive transport mechanism which involves an enzyme as the carrier. Verification must await the isolation and full characterization of the specific enzyme(s) involved in the transport of a given molecule. Transport of cations by membrane diffusable macrocyclic antibiotics (valinomycin, nigericin, for example) also belongs to the category of carrier mediated passive

transport. Synthetic macrocyclic compounds (crown ethers, crypt-
ands, for example) are increasingly utilized for obtaining fundamental understanding of carrier mediated transport mechanisms in
membrane models.

2) Channel mediated transport - cations are mainly transported by their passive diffusion through channels (or pores) in
the membranes. Gramicidin A is the best understood channel forming
substance. It is a linear polypeptide constituted from 15 neutral
amino acids. Two molecules of Gramicidin A reversibly associate to
form a head to head dimer which spans approximately 30 Å, the
thickness of a typical membrane, Figure 4 ($\underline{4}$). Conductance
measurements across a Gramicidin A containing membrane (at a fixed
potential) result in small positive current jumps of constant
amplitude which correspond to the association and dissociation of
the dimers and, hence, to the opening and closing of the ion
channels. Gramicidin A ceases to facilitate the transport of
cations in membranes thicker than 30 Å. Apparently, the channel
forming dimers do not span thick membranes. Conversely, the
ability of valinomycin to transport cations does not diminish in
thick membranes. These observations are in accord with Gramicidin
A forming channels of defined lengths and valinomycin acting as a
diffusable carrier in the membrane.

3) Gate mediated transport - anions are mainly transported
by their facilitated diffusion through a swinging gate formed by a
transmembrane enzyme undergoing conformational changes, Figure 5.
Exchange of HCO_3^- for Cl^- through the erythrocite membrane during
the flow of blood is believed to occur through this mechanism.

Transport by flux-coupling (co-transport or symport):
Enhanced permeability of a molecule in the presence of another has
been observed. For example, in some membranes the transport of
D-glucose (but not L-glucose!) is substantially increased by the
presence of sodium ions. The enhanced transport is the consequence
of having more than one recognition site on a given transport
protein. Sodium ions bind complimentarily to the glucose transporting enzyme and, hence, facilitate its passage across the
membrane.

Active Transport. By definition, active transport occurs in the
absence of any electrochemical potential originating in a concentration gradient ($\underline{4,6}$). Active transport is driven by a coupled
chemical reaction. Distinction is made between primary and secondary active transport.

Primary active transport: Primary active transport is quite
simply the coupling of a local chemical reaction ($X \longrightarrow Y$) to provide
energy for an uphill facilitated (by E, which may be a carrier, a
channel or a gate) diffusion of a species S across the membrane:

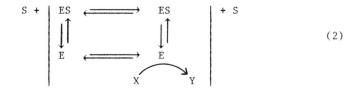

(2)

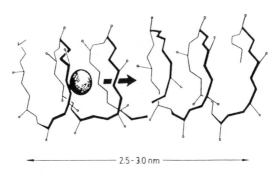

Figure 4. Projection of a three-dimensional model of an electrically conducting pore of gramicidin A. To span the full thickness of the lipid bilayer membrane, two molecules, end-to-end, are required. The side chains of the amino acids are not shown. The model was originally proposed by Urry Proc. Nat. Acad. Sci. USA **68**, 672 (1971). Reproduced with permission from Ref. 4. Copyright 1983, Springer-Verlag.

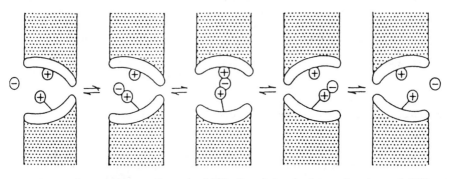

Figure 5. Diagram of a simplified model of the mechanism of Cl⁻ exchange diffusion through a nonconducting pore of the erythrocyte membrane. The gate mechanism is shown functioning in combination with a conformational change in the pore wall. The basic concept is that the gate can only flip over from the cis to the trans position and back if a chloride ion is bound. A conformational change then takes place nearby in the protein, which leads to a screening of the binding site from the cis side and an opening towards the trans side. For simplicity, the conformational change shown in the diagram affects the whole protein. Reproduced with permission from Ref. 4. Copyright 1983, Springer-Verlag.

Energy is provided, for example, by ATP for pumping sodium ions out of and potassium ions into the cell. Another important example of primary active transport is the proton concentration gradient driven ATP synthesis (Mitchell-hypothesis).

Secondary active transport: Secondary active transport is more complex. It involves the permeation of two different substances (A and B) across the membrane. The transport of A is active - it is an uphill process driven by the chemical reaction X→Y. The transport of B is passive, but facilitated by a carrier C, which co-transports A (Equation 3). Co-transport is defined above in the section on passive transport.

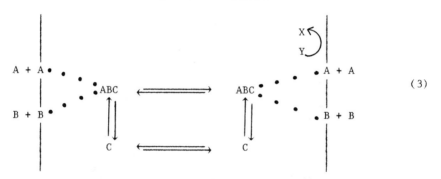

$$(3)$$

Isotonic water resorption in the ephitelium is an example for the secondary active transport. Water and sodium ions are symported from the blood isotonically (i.e., against their concentration gradients) and there is no transport of either in the absence of the other.

Equations have been recently derived for a generalized scheme encompassing primary and secondary active transport systems (7).

Membrane Mimetic Chemistry

Membrane mimetic chemistry is a rapidly emerging discipline concerned with the development of processes which are inspired by the biological membrane (8). Surfactant aggregates - micelles, monolayers, organized multilayers (Langmuir-Blodgett films), bilayer lipid membranes (BLMs), vesicles and polymerized vesicles have been used as media in membrane mimetic chemistry. Different aggregates formed from surfactants are illustrated in Figure 6.

Aqueous micelles are 40-80 Å diameter spherical aggregates which are dynamically formed from surfactants in water above a characteristic concentration, the CMC (9). Depending on the chemical structure of their hydrophilic headgroups, surfactants can be neutral or charged (positively or negatively). The alkyl chain of the surfactants typically contains between 5-20 carbon atoms. Micelles rapidly break up and reform by two known processes. The first process occurs on the microsecond time scale and is due to the release and subsequent reincorporation of a single surfactant from and back to the micelle. The second process occurs on the millisecond time scale and is ascribed to the dissolution of the

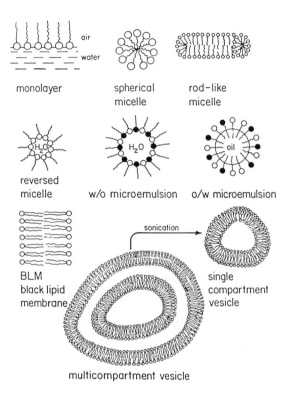

Figure 6.   An oversimplified representation of organized aggregates formed from surfactants.

micelle and to the subsequent reassociation of the monomers.
Substrate interaction with the micelle is also dynamic.

Monolayers (monomolecular layers) are formed by spreading
naturally occurring lipids or synthetic surfactants, dissolved in a
volatile solvent, over water in a Langmuir trough (10). The polar
headgroups of the surfactants are in contact with water, the
subphase, while their hydrocarbon tails protrude above it. Monolayers are characterized by surface area - surface pressure curves,
surface potentials, and surface viscosities. In the gaseous state,
surfactants float freely, mostly lying flat, on the surface without
exerting much force on each other. Monolayers in their gaseous
state may be infinitely expanded without any phase change. Com-
pressing the gaseous monolayers results in a transition to a fluid
state. At least two fluid subphases have been recognized. The
initial transition on decreasing the surface area of gaseous
monolayers results from a gradual reorganization of molecules to a
position more or less perpendicular to the subphase surface. In
this state, the average intermolecular distances are much greater
than that in bulk liquids. On further compression, the distance
between the surfactant headgroups decreases and the system assumes
the liquid condensed fluid phase. In the solid phase, surfactants
in the monolayer are packed as closely as possible; they all are
perpendicular to the subphase or are tilted at an angle. Monolayers in their solid phase show low compressibility as indicated
by the vertical surface pressure-surface area isotherm (Figure 7).
Ultimately, compression leads to a break or inflection in the
isotherm which corresponds to the collapse of the monolayer into
bilayers and multilayers.

Techniques have been developed for transferring the monolayer
onto a solid support and for building up organized multilayer
assemblies in controlled topological arrangements (Figure 8) (11).
Depending on the monolayer forming material and on the mode of
deposition, three structurally different multilayers are recognized. The X-type multilayers (plate-surfactant tail-surfactant
head-tail-head, etc.) are formed by the sequential hydrophobic
attachments of monolayers onto the plate upon immersion only. The
Y-type multilayers (plate-surfactant tail-surfactant head-head-
tail-tail, etc.) are built up both by dipping and by withdrawing
the plate through the floating monolayer. The Z-type multilayers
(plate-surfactant head-surfactant tail, head-tail-head, etc.) are
the result of sequential hydrophilic attachments of the monolayers
onto the plate upon withdrawal only. Absolute and scrupulous
cleanliness is a must in all monolayer and multilayer studies.
Monolayers and multilayers have been stabilized by polymerization
(12-14).

Bilayer (black) lipid membranes, BLMs, are formed by brushing
an organic solution of a surfactant (or lipid) across a pinhole
(2-4 mm diameter) separating two aqueous phases (15,16). Alternatively, BLMs can be formed from monolayers by the Montal-Mueller
method (17,18). In this method, the surfactant, dissolved in an
apolar solvent, is spread on the water surface to form a monolayer
below the teflon partitioning which contains the pinhole (0.1-0.5
mm diameter). Careful injection of an appropriate electrolyte
solution below the surface raises the water level above the pinhole

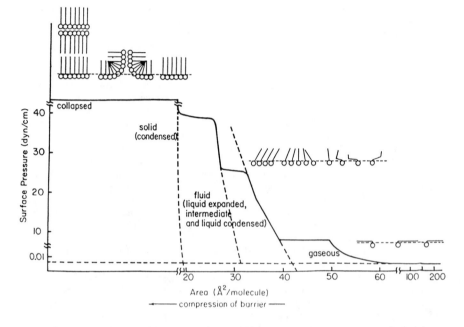

Figure 7.   Schematic representation of a surface pressure –
surface area isotherm for monolayers.

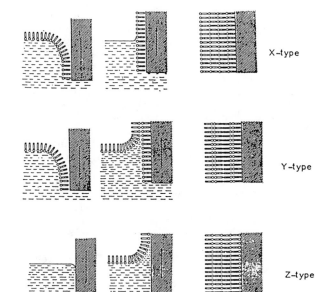

Figure 8. Types of monolayer deposition and resulting system if no rearrangement occurs.

and brings the monolayer into apposition to form the BLM. An advantage of the Montal-Mueller method is that it permits the formation of disymmetrical BLMs. The initially formed film is rather thick and reflects white light with a grey color. Within a few minutes the film thins and the reflected light exhibits interference colors that ultimately turn black. At that point the film is considered to be bimolecular (40-60 Å, thickness). BLMs have been extensively utilized in the elucidation of transport mechanisms by electrical measurements.

Vesicles are smectic mesophases of surfactants containing water between their bilayers (19). Prepared by sonication from such simple surfactants as dioctadecyldimethylammonium bromide (DODAB) or dihexadecylphosphate (DHP), they are single bilayer spherical aggregates with diameters of 500-1000 Å and bilayer thickness of ca. 50 Å. Once formed, vesicles, unlike micelles, do not break down on dilution. Nevertheless, they are dynamic structures. They undergo phase transition, fuse, and are osmotically active. Molecular motions of the individual surfactants in the vesicles involve rotations, kink formation, lateral diffusion on the vesicle plane, and transfer from one interface of the bilayer to the other (flip-flop). Vesicles are capable of organizing a large number of molecules in their compartments. Hydrophobic molecules can be distributed among the hydrocarbon bilayers of vesicles. Polar molecules may move about relatively freely in vesicle-entrapped water pools, particularly if they are electrostatically repelled from the inner surface. Small charged ions can be electrostatically attached to the oppositely charged vesicle surfaces. Species having charges identical with those of the vesicles can be anchored onto the vesicle surface by a long hydrocarbon tail.

The need for increased stabilities, controllable sizes, and permeabilities led to the development of polymerized surfactant vesicles (12-14,20). Vesicle-forming surfactants have been functionalized by vinyl, methacrylate, diacetylene, isocyano, and styrene groups in their hydrocarbon chains or at their headgroups. Accordingly, surfactant vesicles could be polymerized in their bilayers or across their headgroups. In the latter case, either the outer or the inner vesicle surfaces could be linked separately (Figure 9). All polymerized vesicles show appreciable stabilities compared with their unpolymerized counterparts. They have extensive shelf lives and remain unaffected by the addition of up to 30% methanol.

Substrate organization in membrane mimetic systems leads to altered solvation, ionization and reduction potentials and, hence, to altered reaction rates, paths and stereochemistries. These properties have been advantageously exploited, in turn, for reactivity control, catalysis, drug delivery and artificial photosynthesis (8). There are only limited examples of the utilization of membrane mimetic systems in permeability control. In order to gain insight into this important area, we have initiated a research program in BLMs. A status report of our activities in this area will be summarized in the next section.

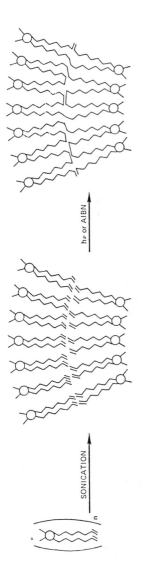
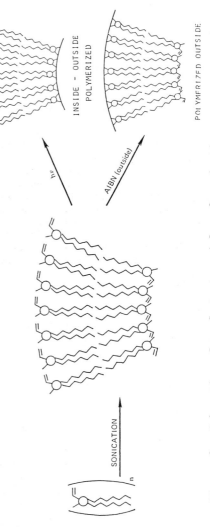

Figure 9.   Schematics of surfactant vesicle polymerization.

Simultaneous Electrical and Spectroscopic Measurements of BLMs

BLMs prepared from phospholipids have been fruitfully utilized in
the past several years in electrical measurements both in the
absence and in the presence of ionophores (21). Holding the
bilayer membrane at a predetermined potential and measuring the
corresponding current flow, i.e., voltage clamping, has contri-
buted much to the present day understanding of ion channels and
impulse transmission (22).
     Investigations of BLMs suffer from two major drawbacks.
First, BLMs are notoriously unstable. Very rarely do they survive
longer than a couple of hours. Second, voltage clamping provides
information only on the transition from an open state to a closed
state in ion channels. Current research in our laboratories is
directed to overcoming these disadvantages by stabilizing BLMs by
polymerization or by polymer coating, and by developing simultaneous in situ spectroscopic and electrical techniques for
monitoring functioning BLMs.
     Direct spectroscopic measurements of absorptions could provide
substantial and much-needed complimentary information on the
properties of BLMs. Difficulties of spectroscopic techniques lie
in the extreme thinness of the BLM; absorbances of relatively few
molecules need to be determined. We have overcome this difficulty
by Intracavity Laser Absorption Spectroscopic (ICLAS) measurements.
Absorbances in ICLAS are determined as intracavity optical losses
(23). Sensitivity enhancements originate in the multipass,
threshold and mode competition effects. Enhancement factor as high
as 10^6 has been reported for species whose absorbances are narrow
compared to spectral profile of the laser (10). The enhancement
factor for broad-band absorbers, used in our work, is much smaller.
Thus, for BLM-incorporated chlorophyll-a, we observed an enhancement factor of 10^3 and reported sensitivities for absorbances in
the order of 10^{-6} (24).
     Figure 10 shows the schematics of the experimental setup used
for intracavity laser absorption spectroscopy (ICLAS) of bilayer
lipid membranes (BLMs). Simultaneous electrical and ICLAS measurements were carried out in a two-compartment container constructed
from two 1 cm path lengths quartz cells (Figure 11).
     ICLAS offered a convenient monitoring of BLM formation. The
upper part of Figure 12 shows the time dependent change of the
relative laser intensity paralleling BLM formation in the cavity.
BLM-forming solution was brushed across the teflon aperture at t =
0. Due to the scattering of the very thick film, initially
present, as well as to non-uniform, large losses in the cavity, no
lasing was observed. After some time, indicated by A in the upper
part of Figure 12 (typically 3-4 minutes), the film sufficiently
thinned, and lasing was observed. Further thinning resulted in a
gradual increase of the transmitted light intensity until it
reached a plateau value (indicated by B in the upper part of Figure
12). At this plateau, true bimolecular thick membranes (BLMs) were
present. The plateau value remained constant until the membrane was
broken (indicated by C in the upper part of Figure 12).
     BLM formation was simultaneously observed by electrical
measurements (see lower part of of Figure 12). A triangular

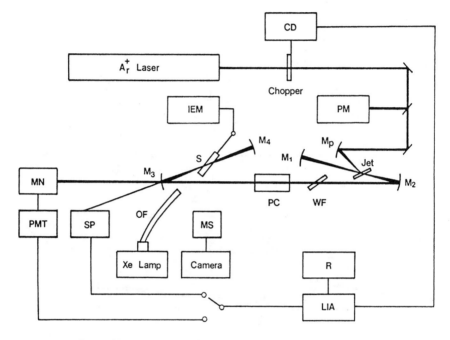

Figure 10.   Schematics of the experimental setup for intracavity
laser absorption spectroscopy (ICLAS). CD = chopper driver; PM
= power meter; M_1, M_2, M_3, M_4 = spherical high reflection
mirrors; M_p = pump mirror; MN = monochromator; PMT = photo-
multiplier; SP = silicon photocell; PC = Pockels cell; WF =
wedged filter; LIA = lock-in amplifier; R = recorder; MS =
microscope; OF = optical fiber; S = sample (solution on BLM);
IEM = instruments for electrical measurements (see Figure 2).

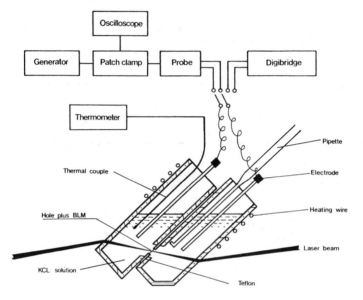

Figure 11.   Schematics for simultaneous ICLAS and electrical measurements of BLM.

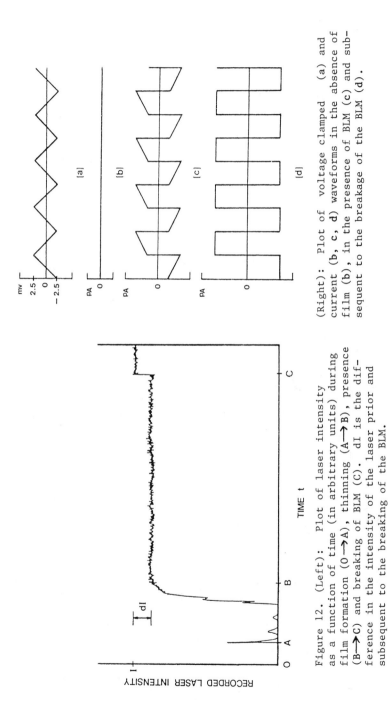

(Right): Plot of voltage clamped (a) and current (b, c, d) waveforms in the absence of film (b), in the presence of BLM (c) and subsequent to the breakage of the BLM (d).

Figure 12. (Left): Plot of laser intensity as a function of time (in arbitrary units) during film formation (0→A), thinning (A→B), presence (B→C) and breaking of BLM (C). dI is the difference in the intensity of the laser prior and subsequent to the breaking of the BLM.

voltage clamped waveform (**a** in the lower part of Figure 12) was
applied across the film. The observed current waveform changed
with the formation of a thick film subsequent to the brushing of
the membrane-forming solution across the teflon aperture, with the
thinning of the film to BLM, and with the breaking of the BLM.
These electrical changes corresponded to changes observed by ICLAS.
Thus, no current passed across the electrodes prior to appreciable
thinning of the membrane. The observed trace **b** in the lower part of
Figure 12 corresponded to the **0⟶A** time domain (see upper part of
Figure 12) observed by ICLAS. Increase in the transmembrane
current corresponded to the thinning of the film to BLM (see **c** in
the lower part and **A-B** in the upper part of Figure 12). The
current waveform remained stable and unaltered during the presence
of the BLM (see **B⟶C** in the upper part of Figure 12). Breaking of
the BLM was signalled by the appearance of perfect square waves
corresponding to the saturation of the amplifier by large electrode
currents (see **d** in the lower part and **C** in the upper part of Figure
12).

Thinning of the film was also observed by microscopy. The
initially white film gradually changed color and showed a variety
of interference fringes (between points **A** and **B** in Figure 12),
which ultimately turned black (at point **B**). Generally, BLM
formation was complete within 20 minutes. Typically, BLMs lasted
for 1-3 hours.

Microscopic observations afforded the calculation of the
physical area of BLM, which, in combination with electrical
measurements, led to values of BLM capacitances per unit area.
Typical BLMs prepared from DODAC (both in the presence and in the
absence of chlorophyll-a) had areas of 5.7×10^{-3} cm^2 and 0.7
$\mu F/cm^2$ capacitances. These values agreed well with those determined
for BLMs prepared from phospholipids (capacitance = 0.7-1.3 $\mu F/cm^2$)
and from single-chain surfactants (capacitance = 0.3-0.6 $\mu F/cm^2$).

Thickness of the insulating layer, d, in DODAC BLMs can be
assessed from:

$$d = \frac{\varepsilon_o \varepsilon A_m}{C} \tag{4}$$

where ε_o is the dielectric constant in vacuum, and taken to be 8.85
$\times 10^{-12}$ CV^{-1} m^{-1}, ε is the dielectric constant of the hydrocarbon
and is assumed to be 2.1 ([15]). A_m is the area of membrane, deter-
mined here to be 5.7×10^{-3} cm^2 and C is the capacitance of the
BLM, determined here to be 4.0 nF. Substituting these values into
Equation 11 gave d = 26.5 Å for the thickness of the insulating
layer in DODAC BLMs. This value is in very good agreement with
those calculated for phospholipid bilayer membranes (23-26 Å) ([25])
making the same assumptions as used here.

We have also prepared BLMs from polymerizable surfactants and
polymerized them in situ ([26]). Extents of polymerization have been
followed by nanosecond, time-resolved fluorescence spectroscopy and
anisotropic measurements ([26]). Experiments have been initiated for
realizing the different biological transport mechanisms in polymerized and partially-polymerized BLMs and for studying their mechanisms by simultaneous electrical and spectroscopic measurements.

Acknowledgments

I thank my co-workers, whose names appear in the references listed, for their enthusiastic, dedicated, and skillful work. The National Science Foundation, Department of Energy, and Army Research Office provided financial support for different aspects of our researches.

Literature Cited

1. Hinze, W. L. in Solution Chemistry of Surfactants, Mittal, K. L., Ed.; Plenum Press: New York, 1979; pp. 79-127.
2. Armstrong, D. W. "Separation and Purification Methods"; 1985, 14, 213-304.
3. Singer, S. J.; Nicolson, G. L. "The Fluid Mosaic Model of the Structure of Cell Membranes"; Science 1972, 173, 720-731.
4. Hoppe, W.; Lohmann, W.; Markl, H.; Ziegler, H. in Biophysics, Springer Verlag: Berlin, 1983.
5. Luzatti, W. "X-ray Diffraction Studies of Lipid-water Systems" (Biological Membranes, Physical Fact and Function), Chapman, C., Ed.; Academic Press: New York, 1968; pp. 71-123.
6. Ovchinnikov, Y. A. in Biochemistry of Membrane Transport, Semenza and Carafoli, Eds., Springer-Verlag: Berlin, 1977.
7. Goddard, J. D. "A Fundamental Model for Carrier Mediated Energy Transduction in Membranes"; J. Phys. Chem. 1985, 89, 1825-1832.
8. Fendler, J. H. "Membrane Mimetic Chemistry"; John Wiley: New York, 1982.
9. Wennerstrom, H.; Lindman, B. "Micelles, Physical Chemistry of Surfactant Association"; Phys. Rep. 1979, 52, 1-86.
10. Gaines, G. L., Jr. "Insoluble Monolayers at Liquid-Gas Interfaces"; Interscience: New York, 1966.
11. Kuhn, H.; Möbius, D. "Systems of Monomolecular Layers - Assembling and Physico-Chemical Behavior"; Angew. Chem. Int. Ed. Engl. 1971, 10, 620-637.
12. Fendler, J. H. "Polymerized Surfactant Aggregates" (Surfactants in Solution), Mittal, K. L. and Lindman, B., Eds.; Plenum Press: New York, 1984, pp. 1947-89.
13. Paleos, C. M. "Polymerization in Organized Systems"; Chem. Soc. Revs. 1985, 14, 45-67.
14. Bader, H.; Dorn, K.; Hashimoto, K.; Hupfer, B.; Petropoulos, J. H; Ringsdorf, H.; Sumimoto, H. "Polymeric Monolayers and Liposomes as Models for Biomembranes" (Polymer Membranes), Gordon, M., Ed.; Springer Verlag: Berlin, 1985, pp. 1-62.
15. Tien, H. T. "Bilayer Lipid Membranes (BLM), Theory and Practice"; Marcel Dekker: New York, 1974.
16. Tien, H. T. in Membranes and Transport, Martonozi, A. N., Ed.; Plenum Press: New York, 1982, p. 165.
17. Montal, M.; Mueller, P. "Formation of Bimolecular Membranes from Lipid Monolyers and a Study of their Electrical Properties"; Proc. Natl. Acad. Sci. USA 1976, 69, 3561.
18. White, S. H.; Petersen, D. C.; Simon, S.; Yafuso, M. "Formation of Planar Bilayer Membranes from Lipid Monolayers. A Critique"; Biophysical J. 1976, 16, 481-489.

19. Fendler, J. H.   "Surfactant Vesicles as Membrane Mimetic
    Agents: Characterization and Utilization"; Acc. Chem. Res.
    1980, 13, 7-13.
20. Fendler, J. H.; Tundo, P.   "Polymerized Surfactant Aggregates:
    Characteriation and Utilization"; Acc. Chem. Res. 1984, 17,
    3-8.
21. Tien, H. T.   "Bilayer Lipid Membranes (BLM).   Theory and
    Practice"; Marcel Dekker: New York, 1974.
22. Hille, B.   "Ionic Channels of Excitable Membranes"; Sinaver
    Associates, Inc.: Sunderland, Massachusetts, 1984.
23. Harris, T. D. in Ultrasensitive Laser Spectroscopy, Kliger,
    David S., Ed.; Academic Press: New York, London, 1983, p.
    343.
24. Zhao, X.-K.; Fendler, J. H.   Submitted for publication, 1986.
25. Alvarez, O.; Latorre, R.   Biophys. J., 1978, 21, 1.
26. Rolandi, R.; Flom, S.; Dillon, I.; Zhao, X.-K.; Fendler, J. H.
    Unpublished work, 1986.

RECEIVED October 24, 1986

Chapter 3

Chromatographic Capabilities of Micellar Mobile Phases

John G. Dorsey

Department of Chemistry, University of Florida, Gainesville, FL 32611

The role of micellar mobile phases is moving from laboratory curiosity to practical utility. The driving force behind the continued interest in these mobile phases lies in the unique chromatographic capabilities they provide. Chemical properties of micelles coupled with the unchanging bulk solvent composition provide the analyst with capabilities unavailable with traditional hydroorganic mobile phases. These include reversed phase gradient elution separations with no column reequilibration, and gradient compatability with electrochemical detection. There has been some disagreement in published work, however, about the efficiency achievable with micellar mobile phases, and about schemes to improve the inherently low efficiency obtained. We review the chromatographic capabilities and present a reexamination of the efficiency problem, and show that with careful attention to mobile phase conditions, efficiencies equivalent to hydroorganic mobile phases are achievable.

In 1980 Armstrong and Henry first effectively demonstrated the usefulness of micellar mobile phases for reversed phase liquid chromatography (1). Since that time several other academic groups have become active in the investigation of these unique mobile phase systems, and the last three years have seen many advances in this area. Yet in spite of the fervor with which micellar chromatography has been promoted by its practitioners, it still has not achieved widespread usage or respect among academic or practicing chromatographers. An interesting perspective on the view of micellar chromatography comes from the 1982 and 1986 Fundamental Reviews issues of ANALYTICAL CHEMISTRY. In 1982 it was said (2):

"An interesting variation on the RPLC/BPLC experiment has been put forth by Armstrong. The

0097-6156/87/0342-0105$06.00/0
© 1987 American Chemical Society

addition of surfactants to the mobile phase above
the critical micelle concentration can have dramatic
effects on the retention behavior of solutes...There
is indication that this method may provide the
possibility for deriving micelle-water equilibrium
constants in an experimentally simpler manner than
previously. It also may provide the ability to make
retention measurements independent of the column
type or manufacturer, and perhaps even a rational
scheme for an 'index' system uncomplicated by varia-
bility in the reversed phase packings. Future developments in 'pseudophase' chromatography will bear
watching."

It was then recognized early in the development of the technique that there were possibilities for dramatic differences in the
chromatographic performance of hydroorganic and micellar mobile
phases. Since that review appeared there have in fact been several
examples of micellar mobile phases providing solutions to inherent
limitations of hydroorganic mobile phases; allowing chromatographic
capabilities that are not possible with traditional mobile phases.
Yet in spite of these advances it was said in 1986 (3):

"The use of substances that form micelles as mobile
phase additives continues to serve as an area of
academic and practical interest. Often touted as a
new form of chromatography, micelle chromatography
should perhaps be considered as a fascinating
example of the incorporation of secondary equilibria
for the enhancement of selectivity and the adjustment of retention. In terms of practical chromatography, it is not yet clear that micelle chromatography solves any problems that cannot be solved by
conventional means. What is more clear is that
micelle chromatography may provide a new route to
the study of micelle phenomena."

In liquid chromatography, the "primary" equilibrium (or quasi-
equilibrium) is the distribution of the solute between the mobile
and stationary phases. Any other equilibria which occur in the
mobile phase, stationary phase, or both are considered
"secondary". Within this rigorous definition, micellar chromatography is indeed an example of secondary equilibria, and like other
secondary equilibria such as acid-base equilibria and ion-pairing
methods, can be used to provide unique chromatographic selectivities
for "difficult" separations. However, unlike other secondary
equilibria methods that are applicable to only narrow ranges of
compounds, micellar chromatography is applicable to a very wide
range of compounds, with the only requirement being that the solute
partition to the micelle. This means that all hydrophobic
compounds, and many hydrophilic compounds which are
electrostatically attracted to the micelle structure, are candidates
for separation by micellar chromatography. As reversed phase is
generally the liquid chromatographic method of choice for hydropho-

bic compounds, this means that micellar chromatography is potentially applicable to a very large percentage of reversed phase separations.

In many forms of secondary equilibria separations, the concentration of the equilibrant, or the mobile phase component which participates in the secondary equilibria, controls, at least partially, the strength and selectivity of the mobile phase. In micellar chromatography the concentration of micelles plays this role, which means that for all separations carried out with micellar mobile phases, the **strength of the mobile phase can be changed while maintaining an unchanging bulk solvent composition.** This unique aspect of micellar mobile phases does indeed allow the solution to "problems that cannot be solved by other means".

The solution to the inherent limitations of hydroorganic mobile phases is in fact the driving force behind the continued interest in micellar mobile phases. Since the first publication on micellar chromatography, the advantages of low cost, low toxicity and chromatographic selectivity have been promulgated. These are not, however, compelling reasons for the practicing analyst to adopt a technique that requires a new learning curve. The adaptation of micellar mobile phases as a routine chromatographic technique will occur because of chromatographic capabilities that are not available with hyroorganic mobile phases. It is these unique chromatographic capabilities that we have been investigating.

Gradient Capabilities

The first chromatographic capability of micellar mobile phases that was shown is the ability to perform reversed phase gradient elution separations with no column reequilibration necessary between samples (4,5). Gradient elution techniques are the most common solution to the general elution problem in liquid chromatography. Snyder has thoroughly addressed the theory of gradient elution, and has shown the advantages of faster separation, higher sample capacity, and lower limits of detection as compared with an isocratic separation (6). However, these techniques have never enjoyed the popularity, especially for routine, repetitive analyses, which would be commensurate with the advantages offered. The sole reason for this is the lengthy column reequilibration necessary after a hydroorganic gradient separation. Reversed phase stationary phases are selectively solvated by the organic component of the mobile phase and the extent of this solvation is dependent upon the composition of the mobile phase. The solvation structure then changes during a gradient elution program, and therefore the column must be 'reequilibrated' with the original (weak) mobile phase. A rule-of-thumb is that up to 20 column volumes of the original mobile phase may be necessary for this reequilibration process. This means that although the **separation** may be speeded by the gradient, the **analysis time,** defined as the time between injections, will often be nearly equivalent of an isocratic separation.

It is well known from ion pairing chromatography that surfactants adsorb onto reversed phase stationary phases. Knowledge of the lengthy equilibrations necessary before ion pairing separations leads to an intuitive belief that gradient elution micellar

chromatography would be futile. However, a unique property of
micelle solutions allows the micelle concentration to be changed
without affecting the structure or composition of the stationary
phase. Micellar aggregates are in dynamic equilibrium with free
surfactant (monomer) and above the critical micelle concentration
(CMC) there is an approximately constant concentration of free surfactant. As it is the free surfactant that interacts with the stationary phase, this means that after an initial equilibration with
any surfactant concentration above the CMC there is no further
change in the amount of adsorbed surfactant. We have two pieces of
chromatographic evidence that this is so.

After a gradient elution separation, failure to fully reequili-
brate the column leads to irreproducible retention of early eluting
compounds. The retention of phenol, which had a k' (capacity
factor) value of 3.05 in 0.05 M sodium dodecyl sulfate (SDS), was
repetitively measured after a gradient program from 0.05 to 0.20 M
SDS and subsequent return to the initial conditions. Only the
volume of initial mobile phase necessary to sweep the injector and
other pre-column volumes was pumped before injection of the phenol
sample. Ten repetitive trials of the gradient and subsequent step
back to initial conditions gave a mean k' value of 3.02 with a relative standard deviation of 0.6% (4). The capacity factor was then
statistically equivalent for both the isocratic separation and following the step back after the gradient, proving that no column
modification occurred during the gradient.

The adsorption isotherm of SDS on a C_{18} stationary phase was
also measured by determining the amount of surfactant adsorbed onto
the stationary phase from frontal chromatography experiments (5).
Figure 1 is a log-log plot of surface concentration vs. mobile phase
concentration of SDS with a standard mobile phase of n-propanol:
water (3:97)(vida infra). The maximum concentration of surfactant
adsorbed on the stationary phase occurs at the mobile phase
concentration of ca. 10^{-2} M and gives a surface concentration of ca.
1.8 μmoles/m^2 of adsorbed SDS. Figure 1 is then supporting
evidence for the conclusion that no column reequilibration is necessary after a micelle concentration gradient. In fact, this plot
should show a break at the CMC value of the surfactant, as that
represents the maximum concentration of free surfactant that will
exist in solution. Because of the nature of the curvature of these
plots, they are not true Langmuir isotherms. That is, they do not
show a break when the stationary phase becomes truly saturated,
rather the break is a result of the micellization of the surfactant.

This advance should then finally allow RPLC gradient techniques
to be useful for repetitive, routine analyses with dramatic saving
of both time and solvent.

Gradient Elution with Electrochemical Detection

To many analysts the major limitation of electrochemical detection
for liquid chromatography (LCEC) is its limited applicability to
gradient elution techniques. Amperometric electrochemical detectors
exhibit both the best and the worst characteristics of solute property and bulk property detectors. While the Faradaic current
arises only from the solute, the non-Faradaic current arises from

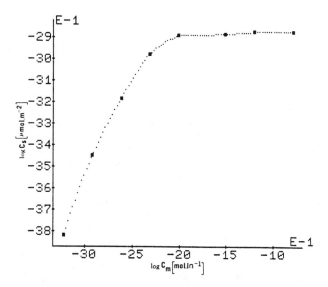

Figure 1. Adsorption isotherm of SDS on an Altex Ultrasphere ODS
column at 30° C.   Mobile phase is n-propanol:water
(3:97).   Apparent saturation of stationary phase is
obvious, but see text.   "Reproduced with permission
from ref. 5. Copyright 1984 Elsevier Science Publishers."

the bulk mobile phase components. This means that while the detector response is solute dependent, and therefore selective, the noise (residual current) is controlled by the mobile phase. During a gradient elution, the composition of the mobile phase changes dramatically and with it the residual current.

We have investigated the extent of baseline shift for both hydroorganic and micellar gradients under different conditions (7). The major contributors to baseline shift were found to be applied potential, cell design, conductance and pH of the mobile phase. While the cell design and applied potential remain constant during a gradient experiment, the magnitude of the applied potential greatly exacerbates the problem of changing mobile phase conductance and pH. At high applied potentials, such as + 1.2V, buffering the solution and balancing the conductance over the gradient range is of critical importance in achieving an "acceptable" baseline shift. However, this cannot be achieved over the entire range of a water to organic gradient. The aqueous buffer systems employed are not totally operative in hydroorganic mixtures and the specific conductance of water-methanol mixtures passes through a minimum, corresponding to maximum viscosity, which makes it virtually impossible to balance the conductance of hydroorganic mixtures. Gradient elution LCEC at high applied potentials can then only be performed over narrow gradient ranges with hydroorganic mobile phases.

Micellar concentration gradients, however, change the bulk properties of the mobile phase to a much less extent than does an organic modifier concentration gradient. The bulk solvent, here 97:3 water:n-propanol, remains constant during a micelle concentration gradient, which makes the control of such parameters as conductance, pH, and even mobile phase impurities much easier. The conductance of a micellar solution is directly proportional to the concentration of ionic surfactant, and therefore, the conductance change during a micellar gradient can be greatly reduced, or even eliminated, by using different supporting electrolyte concentrations in the two surfactant solutions. For non-ionic surfactants the solution conductivity is totally controlled by the amount of added supporting electrolyte. Aqueous buffers also work well in the presence of a small, but fixed, percentage of organic modifier.

Therefore, micellar concentration gradients allow the control of mobile phase conductance and pH, and are highly compatible with amperometric electrochemical detection. With an applied potential of + 1.2V, a gradient from 0.01 M SDS to 0.40 M SDS resulted in a baseline shift of only 8 nA. Both solutions were buffered at pH 2.35 and the conductivity of the two solutions was balanced from the addition of 0.226 M NaClO$_4$ to the 0.01 M SDS solution and 0.05 M NaClO$_4$ to the 0.40 M SDS solution. It should be emphasized that this is a "worst case" experiment, and that smaller gradient ranges and lower operating potentials would result in even less baseline shift. Figure 2 shows a gradient separation of some phenolic compounds, and while the gradient conditions have not been optimized for the best separation, the possibility of performing gradient elution separations with an electrochemical detector at a high applied potential is clear.

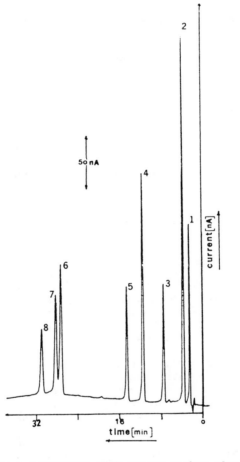

Figure 2. Gradient separation with glassy carbon electrode at
+ 1.2V.   Flow   rate   1.0   mL/min.   Column: Altex
Ultrasphere ODS.   Mobile phase A: 0.05 M SDS, 3% n-
propanol,   pH   2.5 with   phosphate   buffer,   sodium
perchlorate added to balance conductivity with solvent
B.   Mobile phase B: 0.112 M SDS, 3% n-propanol, pH 2.5
with phosphate buffer.   Gradient program A to B in 12
min.   Peak   identification: (1)   hydroquinone;   (2)
resorcinol;   (3)   catechol;   (4)   phenol;   (5)   p-
nitrophenol;   (6)   o-nitrophenol;   (7)   p-chlorophenol;
(8) p-bromophenol.   "Reproduced with permission from
ref. 5. Copyright 1984 Elsevier Science Publishers."

Efficiency Enhancement

A major drawback in the early reports of micellar chromatography was a serious loss of efficiency when compared to traditional hydroorganic mobile phases. If micellar mobile phases are ever to be widely accepted as a viable chromatographic technique, the efficiency achieved must at least approach that of conventional reversed-phase LC.

Care must be taken in the use and interpretation of literature efficiency values. Efficiency values are likely the most incorrectly calculated chromatographic figure of merit. The commonly used equations based on peak width at the base or at half the height of the peak are valid only for perfectly Gaussian shaped peaks. This problem has been realized by chromatographers for some time, but the popularity of these methods continues because until recently the only alternative was computer based moment calculations. Kirkland et al. addressed this problem and recommended that peak symmetry values be reported along with efficiency values (8). They further showed that the calculation based on width at half height can give values as much as 100% in excess of the actual value. Because of this, care must be taken in the use of literature plate count values.

Foley and Dorsey have recently derived a simple manual method for the calculation of plate counts that corrects for the asymmetry of skewed peaks (9). This equation has been used in all of our micellar efficiency calculations and is:

$$N = 41.7(t_r/W_{0.1})^2/(B/A) + 1.25$$

This equation has recently been shown to be the most accurate manual method of plate count calculation (10).

Dorsey et al. were the first to address the low efficiency of micellar mobile phases, and through the use of plots of reduced plate height vs. reduced velocity (Knox plots) they showed the efficiency limiting problem to be poor mass transfer (11). That this is the problem is not surprising. While micellar mobile phases offer unique chromatographic advantages, the separation method is still reversed phase chromatography. Since the invention of bonded reversed phase materials, it has been known that totally aqueous mobile phases will give poor efficiency and peak shape. In 1975 Kirkland said (12):

>   "The bonded hydrocarbon packings...are very hydrophobic...Therefore, in reversed phase separations...it is desirable to use aqueous mobile phases containing ⩾~10% of a miscible organic solvent...to improve wetting characteristics. Mobile phases with no or low concentrations of organic solvent produce broad peaks because of the slow equilibrium resulting from the resistance to solute mass transfer across the interface of the two very unlike phases."

In 1977 Scott and Kucera reported (13):

> "The wetting characteristics can be extremely
> important in the practical uses of reversed
> phases. If the water content is increased...and the
> water present exceeds the wetting limit, a signifi-
> cant interfacial resistance to mass transfer effect
> could be produced, which would severely impair
> column efficiency."

The goal for reversed phase micellar chromatography is then to provide wetting with the necessary organic solvent while perturbing the micelle structure as little as possible. Here again, knowledge of reversed phase techniques is helpful. Scott and Simpson have studied modification of C_{18} phases by organic modifiers and have shown that over 90% of the surface is covered with the alcohol at a concentration of 3% (w/v) n-propanol, but there is only about 50% coverage with the same concentration of methanol (14). This modifi-cation of the surface should then allow fast mass transfer, resulting in improved efficiencies.

We have found that the use of 3% n-propanol in the micellar mobile phase and column temperatures of 40° C appear to offer a broadly applicable solution to the low efficiency previously repor-ted for micellar mobile phases. These conditions have resulted in reduced plate heights of 3-4 for SDS, cetyltrimethylammonium bromide (CTAB), and Brij-35 (15). This efficiency optimization scheme then appears to be a broadly-based solution for micellar mobile phases of any surfactant. This means that the surfactant type can be varied to affect separational selectivity with no loss in column efficiency.

That the significant problem causing low efficiency is poor wetting of the stationary phase has recently been confirmed by Foley and May (16). In a study of optimization of pH for the separation of weak organic acids on hydrophobic stationary phases, they studied column efficiencies with purely aqueous (non-micellar) mobile phases and investigated adding small amounts of methanol, ethanol, n-propa-nol, and acetonitrile as a means of improving efficiency. They found that n-propanol was by far the most effective organic solvent, with 3-6% (v/v) improving chromatographic efficiencies by factors of 10-15, which approached the efficiencies obtained with traditional hydroorganic mobile phases. Furthermore, they observed only slight improvements with the other solvents. This is consistent with pre-vious findings that methanol, ethanol and acetonitrile are ineffec-tive at increasing the efficiency of micellar mobile phases (11).

While the added propanol does somewhat modify the micelle structure, when the goal is chemical analysis, it is necessary to provide efficiencies equivalent to hydroorganic mobile phases, and it still allows the practicing chromatographer to take advantage of the unique capabilities of these mobile phases. As stated in the 1986 Fundamental Review issue of ANALYTICAL CHEMISTRY, micellar chromatography can be used not only for analysis, but also for the study of micelle phenomenon. Here, certainly, chromatographic effi-ciency is not the primary consideration and the added propanol would complicate matters unnecessarily. As shown initially by Armstrong

(17) and later by Arunyanart and Cline Love (18) micellar mobile phases provide an excellent way of obtaining micelle-water partition coefficients. Other fundamental studies will certainly be forthcoming.

Conclusion

Micellar mobile phases will never replace traditional hydroorganic mobile phases. They do, however, deserve serious consideration by practicing chromatographers as they can provide the solution to certain fundamental limitations of hydroorganic mobile phases. Hopefully the advantages will overcome the skepticism and resistance to change shown by many chromatographers and micellar mobile phases will soon assume a role of importance.

Literature Cited

1. Armstrong, D.W.; Henry, S.J. J. Liq. Chromatogr. 1980, 3, 657–662.
2. Majors, R.E.; Barth, H.G.; Lochmüller, C.H. Anal. Chem. 1982, 54, 323R–363R.
3. Barth, H.G.; Barber, W.E.; Lochmüller, C.H.; Majors, R.E.; Regnier, F.E. Anal. Chem. 1986, 58, 211R–250R.
4. Landy, J.S.; Dorsey, J.G. J. Chromatogr. Sci. 1984, 22, 68–70.
5. Dorsey, J.G.; Khaledi, M.G.; Landy, J.S.; Lin, J.-L. J. Chromatogr. 1984, 316, 183–191.
6. Snyder, L.R. in "High Performance Liquid Chromatography: Advances and Perspectives", Horvath, C., Ed.; Academic Press: New York, 1980, Vol. 1.
7. Khaledi, M.G.; Dorsey, J.G. Anal. Chem. 1985, 57, 2190–2196.
8. Kirkland, J.J.; Yau, W.W.; Stoklosa, H.J.; Dilks, C.H., Jr. J. Chromatogr. Sci. 1977, 15, 303–316.
9. Foley, J.P.; Dorsey, J.G. Anal. Chem. 1983, 55, 730–737.
10. Bidlingmeyer, B.A.; Warren, F.V., Jr. Anal. Chem. 1984, 56, 1583A–1596A.
11. Dorsey, J.G.; DeEchegaray, M.T.; Landy, J.S. Anal. Chem. 1983, 55, 924–928.
12. Kirkland, J.J. Chromatographia 1975, 8, 661–668.
13. Scott, R.P.W.; Kucera, P. J. Chromatogr. 1977, 142, 213–232.
14. Scott, R.P.W.; Simpson, C.F. Faraday Symp. Chem. Soc. 1980, 15, 69–82.
15. Landy, J.S.; Dorsey, J.G. Anal. Chim. Acta 1985, 178, 179–188.
16. Foley, J.P.; May, W.E. Pittsburgh Conference and Exposition 1985, Abstract 1207.
17. Armstrong, D.W.; Nome, F. Anal. Chem. 1981, 53, 1662–1666.
18. Arunyanart, M.; Cline Love, L.J. Anal. Chem. 1984, 56, 1557–1561.

RECEIVED October 24, 1986

Chapter 4

High-Performance Liquid Chromatography of Organic and Inorganic Anions: Use of Micellar Mobile Phase

Frank G. P. Mullins

Department of Instrumentation and Analytical Science, University of Manchester
Institute of Science and Technology, Manchester, M60 1QD, United Kingdom

Ion pairing chromatography has been widely used
for the chromatographic determination of ionizable
solutes. To date, the most commonly used ion-
pairing reagents are non-micelle forming, but there
can be advantages in using the micelle forming
reagents. For example, no method existed previously
for the separation and determination of
dithiocarbamate salts, widely used as fungicides.
High-performance liquid chromatography using
micellar hexadecyltrimethylammonium chloride as
the mobile phase provides a versatile and efficient
technique for the separation of iodate, nitrite,
bromide, nitrate and iodide. The distribution and
retention of the inorganic anions is governed by
their partitioning between the micelles and the
mobile phase, and between the conditioned stationary
phase and the mobile phase. Dithiocarbamate
salts of varying hydrophobicity can be separated
using micellar hexadecyltrimethylammonium bromide.
Chromatographic efficiency measurements obtained
for the hydrophobic solutes (phenol and benzene)
and the ionic dithiocarbamate salts show that
efficiency remains high even with high concentrations
of methanol as the mobile phase modifier, with
acetonitrile as the modifier, the efficiency falls as
the concentration increases.

The analysis of non-polar solutes by high performance liquid
chromatography is generally a simple task, especially if
reversed-phase systems are used. However, many compounds of
environmental interest, such as dithiocarbamate salts and inorganic
anions are ionized species. By their very nature it is difficult to
chromatograph these well hydrated hydrophilic species.

0097-6156/87/0342-0115$06.00/0
© 1987 American Chemical Society

A method for extracting ionized solutes into organic phases has been studied for a number of decades. Ions of opposite electrical charge are added to the aqueous phases resulting in ion-pairing between the solute ion and the pairing ion. The resultant complex has a low net electrical charge or polarity, is thus poorly hydrated, and so can partition from the aqueous to the organic phase. Jonkman (1) reviewed the area of bulk-phase extraction of ionized drugs, and gave examples of solute extraction where selectivity and sensitivity of approach could be demonstrated. Higuchi and Michaelis (2) and Modin and Schill (3) have also reported work on extraction techniques and applications.

HPLC has been used for ion-analysis using normal-phase adsorption techniques. This however often results in high solute retention coupled with very poor peak shape and poor solute resolution. High-pressure ion-exchange chromatography has also been used for ion-analysis, but the unfavourably high compressibilities of the materials, e.g. a polystyrene- divinylbenzene matrix cation exchanger, does not permit efficient high speed separations to be made. The requirements of an HPLC method for the analysis of ionized solutes are rapidity, sensitivity, efficiency and an ability to resolve material from complex systems, such as untreated sewage, without prior extraction.

Ion-Pair Chromatography

In cases where the sample is ionizable (e.g. an acid or a base) it is possible to alter the chromatographic retention by introducing long-chain ionic alkyl compounds into the mobile phase. These substances are the types that are used to form classical "ion-pairs" with the sample in a liquid-liquid extraction using a separatory funnel. The addition of these reagents to a liquid chromatographic (LC) eluent will substantially alter the retention of the ionic compounds and will not affect the retention of non-ionic compounds. Because of the similarity of reversed-phase LC to classical liquid-liquid countercurrent extraction, and because of the use of reagents which are similarly used in both classical and LC ion pair extractions, the technique of adding long chain ionic alkyl reagents to a LC eluent has been termed "ion-pair chromatography". However, ion-pair chromatography is a term that describes a chromatographic result (a phenomenon) and not necessarily a cause.

For reversed-phase ion-pair chromatography a non-polar surface (e.g. C_8 or C_{18}) is used as a stationary phase and an ionic alkyl compound is added to the aqueous mobile phase as a modifier. For the separation of acids, an organic base (e.g. tetrabutylammonium phosphate) is added to the eluent; for the separation of bases, an organic acid (e.g. octane sulphonate) is used. Reversed-phase ion pairing is presently the most popular approach because of the simpler technical requirements and very high column performance. It is however essential to operate the system only after equilibrium of the mobile phase and the stationary phase has occurred in order to obtain reproducible analyses.

The application of reversed-phase ion-pair chromatography to the separation of charged solutes has gained wide acceptance mainly because of the limitations of ion-exchange chromatography in

separating both neutral and ionic samples, and because of the
difficulty in separating ionic components by the reverse-phase
techniques of ion-suppression. There have been significant
contributions by a number of authors and several reviews have been
published on ion-pair chromatography on chemically bonded phases
(4, 5). Haney et al. (6, 7) and Knox and Jurand (8) were amongst
the first to develop the technique for widespread use. Knox
continued in the development of the technique, particularly using
long chain hydrophobic counterions. This method advanced rapidly
and has been applied to such diverse areas as peptides and proteins
(9, 10), sulphonated dyes (11), drug substances (12), catecholamines
(13) and alkaloids (14).

Mechanism of Reversed-Phase Ion-Pair Chromatography

Many applications of reversed-phase ion-pair chromatography involve
the addition of long chain alkyl sulphonate ions to the mobile phase
to give enhanced separation of oppositely charged sample ions. This
technique has been called "soap chromatography" (8, 15), "ion-pair
chromatography" (3), "solvent-generated dynamic ion-exchange
chromatography" (16, 17), "hetaeric chromatography" (18),
"detergent-based cation exchange chromatography" (16),
"solvophobic-ion chromatography" (19), and "surfactant
chromatography" (4). The variety of nomenclature indicates the
uncertainty which exists concerning the retention mechanism in this
mode of HPLC.
    There are three popular hypotheses. Two models propose extreme
situations and each encompasses a substantial amount of
chromatographic data. These two proposals are the ion-pair model
and the dynamic ion-exchange model. The third view, which is
broader in scope than the previous two concepts, accommodates both
the extreme views without combining the two models. This proposal
is the ion-interaction model.
    The ion-pair model stipulates that formation of an ion-pair
occurs in the aqueous mobile phase (16, 18, 20). The retention time
is governed by the extraction coefficient of the ion-pair. A longer
alkyl chain on the pairing agent simply makes a less polar ion-pair,
with a resulting higher extraction coefficient, and the retention of
the ion-pair increases as a result of its greater affinity for the
stationary phase.
    The second view stipulates an ion-exchange mechanism (21, 16,
19, 22). In this hypothesis, it is the unpaired hydrophobic alkyl
ions that adsorb onto the non-polar surface and cause the column to
behave as an ion-exchanger. As the chain length of the ion-pairing
reagent increases, the surface coverage of the stationary phase
increases, with a concomitant increase in retention of the ionic
sample.
    The third view, the ion-interaction model, has been proposed by
Bidlingmeyer et al. (23) which is less restrictive than the other
two models previously described. The model is based on conductance
measurements involving neutral and charged samples injected into
solutions containing positively and negatively charged hydrophobic
ions. These measurements show that ion pairs do not form in the
mobile phase. Neither the ion-pairing nor the ion-exchanging model

can explain the data in a consistent way. Instead, the results
suggest a retention mechanism that is broader in scope and is best
described as one of ion-interaction. The ion-interaction mechanism
does not require ion-pair formation in either phase and is not based
on classical ion-exchange chromatography. The ion-interaction
mechanism assumes dynamic equilibrium of the hydrophobic ion
resulting in an electrical double layer forming on the surface. The
retention of the sample results from an electrostatic force due to
the surface charge density provided by the reagent ion, and from an
additional "sorption" effect onto the non-polar surface.

In the ion-interaction model a layer of hydrophobic ions
(ion-pair reagent) is adsorbed onto the non-polar surface. Because
these hydrophobic ions carry the same charge, they are well spaced
from one another, and most of the surface therefore is still the
original non-polar packing surface and only a small amount of the
surface is coated with the reagent. However if the chain length is
significantly long, and its concentration in the mobile phase is
significantly high, it is conceivable to expect the bonded phase to
be coated to a very high level. This would change the nature of the
bonded surface. A primary ion-layer and an oppositely charged
counter-ion layer are formed on the surface of the bonded phase.
This is an electrical double-layer model. Since the adsorbed ions
are in dynamic equilibrium between the bonded phase and the mobile
phase, an increase in the reagent concentration in the mobile phase
leads to an increase in the amount of reagent ion adsorbed, thus
increasing the amount of surface charge. Transfer of samples
through the double layer is a function of electrostatic and Van der
Waals forces. For instance, an ionic organic solute such as
dithiocarbamate anion, is attracted to the charged surface. The
chromatographic retention of the dithiocarbamate results from this
Coulombic attraction and from an additional "sorption" of the
hydrophobic portion of the sample molecule onto the non-polar
surface.

The debate as to the exact model to describe the ion-pair
phenomena will no doubt continue. Difficulties in devising a model
arise from conflicting conclusions based on a large amount of
experimental data. However, it is important to emphasise that
theory guides experimentation. Therefore the importance of having a
model is to understand the factors that control chromatographic
retention, and thus, to aid in the prediction of the separating
ability of a mobile phase.

Micellar Chromatography

Previously most ion-pairing chromatographic separations involved the
use of ion-pairing reagents not capable of forming micelles.
Quaternary ammonium salts containing one long hydrophobic alkyl
chain are called amphiphiles, e.g. hexadecyltrimethylammonium
bromide. These can form micelles in polar solutions, i.e. the
hydrophobic ions interact to form discrete aggregates possessing a
hydrophobic core and a polar surface. Quaternary ammonium salts not
possessing a long hydrophobic alkyl chain such as tetrabutylammonium
bromide are not amphiphiles and cannot form micelles.

A micellar mobile phase differs from a conventional ion-pairing
mobile phase in two important aspects. Firstly, micellar solutions

can be regarded as microscopically heterogeneous, being composed of
the micellar aggregate and the "bulk" surrounding medium. An
ion-pairing mobile phase is homogeneous. Secondly, the
concentration of surfactant in micellar chromatography is above its
critical micelle concentration (CMC), i.e. the concentration above
which micelle formation becomes appreciable. Below its CMC,
hexadecyltrimethylammonium bromide can be used as an ion-pairing
reagent. High performance liquid chromatographic separations
performed with a micellar mobile phase have been reported previously
(24-27).

The separation of anions by the use of a cationic micellar
mobile phase results in a high degree of flexibility not available
from other methods of ion chromatography. The importance of
micelles in the mobile phase lies in their ability to participate in
the partitioning mechanism. The three equilibria involved in
micellar chromatography are schematically represented in Figure 1.
The elution behaviour of the anionic solute depends on three
partition coefficients: K_{mmp}, the partition coefficient between the
bulk mobile phase and and the micelle; K_{bm}, the partition
coefficient between the bonded phase and the micelle and K_{bmp}, the
partition coefficient between the bonded phase and the bulk mobile
phase.

Determination of Inorganic Anions By High Performance Liquid Chromatography

Following the development of ion chromatography by Small et al. (28
considerable interest has been shown in the determination of
inorganic anions by HPLC. In the procedure employed by Small et al.
a column packed with a proprietary anion-exchange resin was
incorporated into commercial instrumentation. Skelly (29) reported
the HPLC separation of inorganic anions using an eluent containing
an octylamine salt. Iskandarani and Pietrzyk (30) and Molnar et al.
(31) demonstrated the determination of anions by using
tetrabutylammonium salts on a styrene-divinylbenzene resin and a
bonded stationary phase, respectively. Cassidy and Elchuk (32, 33)
reported the analysis of inorganic anions using a cyano-bonded
normal-phase column. De Kleijn (35) used hexadecyltrimethylammonium
chloride in order to obtain the separation of inorganic anions
using the same conditions as those employed by Reeve (34).

Determination of Inorganic Anions By High Performance Liquid Chromatography Using a Micellar Mobile Phase

When strongly hydrophobic cationic surfactants are present in the
mobile phase the hydrophobic surface of the stationary phase becomes
dynamically conditioned with respect to the adsorption of the
surfactant. This confers an ion-exchange capability on the
stationary phase. Cassidy and Elchuk (32, 33) reported use of
cetylpyridinium chloride to coat the stationary phase
"permanently", but used tetrabutyl and tetramethylammonium salts in
the mobile phase. Their equilibration procedure also employed the
use of acetonitrile in the initial conditioning step, thus
increasing the overall cost of the analysis. Knox and Hartwick (36)

proposed that retention was a linear function of the charge density
on the surface of the stationary phase. Hung and Taylor (37)
reported adsorption isotherms for hexadecyltrimethylammonium bromide
that indicate a high loading of the stationary phase at 0% organic
modifier. Therefore, the strong retention of anions on a saturated
stationary phase can be attributed to strong anion interaction with
the very high charge density on the surface.

Mullins and Kirkbright (38) reported separation of the UV
absorbing anions; iodate, nitrite, bromide, nitrate and iodide
using a micellar mobile phase containing hexadecyltrimethylammonium
chloride above its CMC. Figure 2 illustrates this separation with
two different concentrations of micellar reagent. Increasing the
concentration of hexadecyltrimethylammonium chloride decreases the
retention time (38) on the column.

The decrease in retention of the anions as the concentration of
hexadecyltrimethylammonium chloride is increased (Figure 2) can be
attributed to anion interaction with micelles in the mobile phase.
The retention time of the selected anions can be reduced by
increasing the concentration of the micellar reagent (Figure 2(a),
(b)). This more rapid analysis time can also be achieved using
organic modifiers such as methanol or acetonitrile with lower
concentrations of micellar reagent. Anion association with a
cationic micelle has been shown to occur in the electric double
layer on the micelle surface (39).

The retention of the anions on a loaded octadecyl-bonded silica
column in the presence of hexadecyltrimethylammonium chloride
micellar mobile phase follows the order

$$I^- > NO_3^- > Br^- > NO_2^- > IO_3^-$$

This order is similar to the anion selectivity order found on a
typical strongly basic anion exchanger. Figure 2(b) illustrates a
separation of five inorganic anions with a $1.36 \times 10^{-1} M$
hexadecyltrimethylammonium chloride micellar mobile phase. No
buffer salts or organic modifier were used in order to accomplish
this separation.

The above results show that conventional HPLC can be used with
UV detection for the determination of inorganic anions, namely IO_3^-,
NO_2^-, Br^-, NO_3^- and I^- with a cationic micellar mobile phase. One
of the attractive features of this procedure is the ability to
control retention by control of the concentration of aqueous
micellar hexadecyltrimethylammonium chloride rather than by the use
of organic modifiers.

Separation of Dithiocarbamates By High
Performance Liquid Chromatography
Using a Micellar Mobile Phase

Dithiocarbamates are those agricultural and horticultural fungicides
which may be considered to be derivatives of dithiocarbamic acids.
Relatively large concentrations of these salts are applied to crops
to achieve adequate disease control. On the crop the
dithiocarbamates may be decomposed photochemically (40), oxidized or
hydrolysed. The decomposition products are more toxic than the
parent dithiocarbamates. Disodium ethylenebisdithiocarbamate is

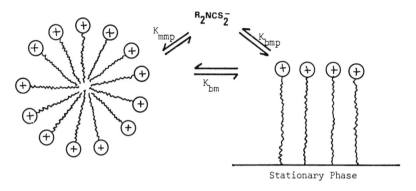

Figure 1. The three equilibria involved in micellar chromatography.

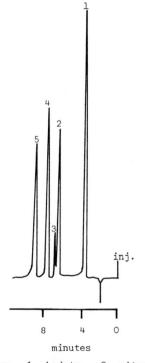

Figure 2. (a) Solutes; 1. iodate; 2. nitrite; 3. bromide;
4. nitrate; 5. iodide; conditions: flow rate, 1.5mL min^{-1};
column packing, ODS Spherisorb; column dimensions, 250 x 5mm;
particle size, 5um; injection volume, 20ul; mobile phase,
1.36 x 10^{-1}M hexadecyltrimethylammonium chloride; detector, UV
photometer at 210nm 0.02 A.U.F.S. "Reproduced with permission
from Ref. 38. Copyright 1984, 'Royal Society of Chemistry,
London'".

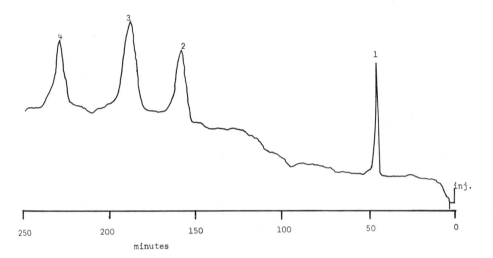

Figure 2. (b) conditions as in (a) but mobile phase. 5 x 10⁻⁴M
hexadecyltrimethylammonium chloride. "Reproduced with permission
from Ref. 38. Copyright 1984, 'Royal Society of Chemistry,
London'".

converted to ethylenethiourea on the crop (41) and also during food
processing e.g. cooking, canning or brewing. Ethylenethiourea has
been reported to produce hepatomas in mice (42) and thyroid
carcinomas in rats (43). Sodium N-methyldithiocarbamate is a soil
fungicide, nematocide and herbicide with a fumigant action applied
at rates of around 11 litres of 32.7% aqueous solution per 100m^2.
The activity of this dithiocarbamate salt is due to its
decomposition to methylisothiocyanate (44).

Determination of Dithiocarbamate Salts by HPLC. Smith et al. (45,
46) reported determination of dithiocarbamate salts by HPLC on an
octadecylsilica column utilizing transition metal salts in the
mobile phase. Mixed ligand formation poses the problem of being
unable to distinguish which dithiocarbamate salt or salts has
chelated to the metal, also the poor detection limit probably makes
this technique unsuitable for the trace analysis of dithiocarbamate
salts. Gustaffson and Thompson (47, 48) reported a procedure for
the determination of dithiocarbamate salts by normal phase HPLC
following extraction and methylation of the salts with methyl
iodide. They report a low recovery, possibly indicating breakdown
of the dithiocarbamate salt during methylation.

Determination of Dithiocarbamate Salts By HPLC Using a Micellar
Mobile Phase. Kirkbright and Mullins (49) reported a
chromatographic technique for separating dithiocarbamate salts based
on the use of micellar hexadecyltrimethylammonium bromide in the
mobile phase. This technique afforded separation of five
dithiocarbamate salts, including disodium ethylene
bisdithiocarbamate in twenty-five minutes on a cyano bonded column.
The separation is illustrated in Figure 3. The micellar mobile
phase also proved to be successful in the separation of sodium
N-methyldithiocarbamate from its decomposition product
methylisothiocyanate (50). The effects of both of the organic
modifiers, methanol and acetonitrile, on the separating ability of
micellar hexadecyltrimethylammonium bromide were also reported (51)
and discussed. The effect of variation of organic modifier
concentration on the efficiency of separations obtained with a
micellar mobile phase has been briefly discussed by a number of
authors. Dorsey et al. (26) advise low concentration of organic
modifier to enhance the mass transfer kinetics of the solute and to
'maintain integrity' of the micelle. Yarmchuk et al. (27) concluded
that the small gains in efficiency were not worth the incorporation
of organic solvents in micellar eluents. Most authors (26, 27) have
used neutral hydrophobic test solutes, which are known to interact
with the hydrophobic core of the micelle, in their efficiency
studies. Yarmchuk et al. (24) discussed the restricted mass
transfer of hydrophobic solutes in micellar chromatography in terms
of the effect of entrance-exit rate constants of phenol and benzene
with micelles. Almgren et al. (52) discussed the dynamic and static
aspects of solubilization of neutral arenes in ionic micellar
systems. They deduced that the exit rates of solutes from micelles
approximately parallel the solubility of the solute in water, i.e.
the greater the solubility of the solute in water, the faster is the
exit rate from a particular micelle. It is proposed from the
deductions of Almgren et al. (52) that the interaction of the

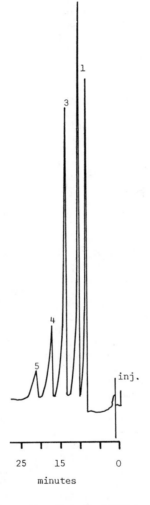

Figure 3. Solutes; 1. sodium N-methyldithiocarbamate;
2. sodium N,N-dimethyldithiocarbamate; 3. ammonium
tetramethylenedithiocarbamate; 4. sodium N,N-diethyldithio-
carbamate; 5. disodium ethylenebisdithiocarbamate. Conditions:
column, 300 x 3.9mm; particle size, 10um; column packing,
u-Bondapak CN; flow rate, 1ml min^{-1}; mobile phase, 1.25 x 10^{-2}M
hexadecyltrimethylammonium bromide, pH 6.8 (phosphate buffer
10mM, 253nm, 0.01 AUFS. "Reproduced with permission from Ref.
49. Copyright 1984, 'Royal Society of Chemistry, London'".

dithiocarbamate salts with the hexadecyltrimethylammonium bromide
micelles must be very rapid because of their high water solubility
e.g. sodium N-methyldithiocarbamate has a water solubility of
$722gl^{-1}$ at 20°C (44). Tagashira (53) discussed the interaction of
dithiocarbamate salts with micelles and proposed that they
interacted with the polar 'mantle' of the micelle.

Conductance measurements of micellar solutions at high
concentrations of methanol and acetonitrile were obtained and
indicate possible rupturing of the "micellar aggregate" in the
acetonitrile/water mobile phase (51). Figure 4 illustrates
separation of four dithiocarbamates, phenol and benzene.

Table I illustrates the difference in the efficiency obtained
with a micellar mobile phase containing (a) methanol and (b)
acetonitrile as the mobile phase modifier.

Table I(a). Variation of Theoretical Plate Number N,
and Resolution, Rs, with Variation in Methanol Concentration

Percent Methanol	Sodium N-methyl-dithiocarbamate			Sodium NN-dimethyl dithiocarbamate			Ammonium tetramethylene-dithiocarbamate		
	k'	N	Rs	k'	N	Rs	k'	N	Rs
30	13.2	2128	6.45	19.9	3025	9.34	31.6	5459	-
50	5.4	3449	4.0	6.7	5352	8.20	10.0	5211	-
70	0.03	3528	1.16	0.5	3595	2.20	2.5	4723	-

Table 1(b). Variation of Capacity Ratio (k') and
Efficiency (N) With Concentration of Acetonitrile

%Aceto-nitrile	Benzene		Phenol		Sodium N-methyl-dithio-carbamate		Sodium N, N-dimethyl dithio-carbamate		Ammonium tetramethylene dithiocarbamate	
	k'	N	k'	N	k'	N	k'	N	k'	N
10	15.0	5436	16.6	3903	23.8	793	-	-	41.2	3281
30	4.0	2740	5.1	2539	5.1	321	6.9	107	11.9	464
50	0.8	228	1.06	124	0.7	90	0.9	23	1.3	86

The efficiency remains high even with high concentrations of
methanol. With acetonitrile as the modifier, the efficiency was
significantly reduced as the concentration of the modifier was
increased. For this study the theoretical plate number (N) was
estimated using the equation of Foley and Dorsey (54).

Conclusions

In conventional reversed-phase ion-chromatography, both the
mobile phase and the stationary phase are chosen to provide the

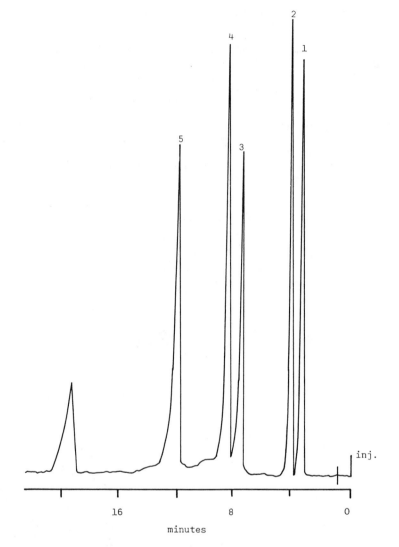

Figure 4. Solutes: 1. benzene; 2. phenol; 3. sodium
N-methyldithiocarbamate; 4. sodium N,N-dimethyldithiocarbamate;
$\overline{5}$. ammonium tetramethylenedithiocarbamate; 6. sodium
diethyldithiocarbamate. Conditions: column, dimensions,
250 x 5mm column packing, Spherisorb ODS; particle size, 5um;
injection volume, 20uL; mobile phase, 1 x 10^{-2}M hexadecyltrimethylammonium bromide, 55% methanol/water; pH 6.8 (phosphate
buffer 10mM); flow rate, 1mL min^{-1}; detection, UV photometer,
254nm (solutes 1, 2), 286nm (solutes 3, 4, 5, 6), 0.01 AUFS.
"Reproduced with permission from Ref. 51. Copyright 1986, 'Royal
Society of Chemistry, London'".

required separation. In micellar chromatography the stationary phase is initially loaded with the surfactant conferring an ion exchange capability to the bonded stationary phase. This is advantageous in that by controlling the concentration of micelles in the mobile phase - and if necessary by careful selection of the organic modifier - the desired loading of the surfactant on the column is controlled, and the optimum separation can be achieved.

Micellar chromatography was applied to the separation of dithiocarbamate salts. Other workers (46) have noted the inadequacy of the ion-pair partition method for the analysis of dithiocarbamate salts, and no alternative method was available that allowed the rapid separation and determination of these salts, commonly used as fungicides. The rapid interaction of organic molecules such as dithiocarbamate salts, possessing a polar functional group and a hydrophobic functional group, with charged micelles is very useful in chromatography. This rapid interaction results in high efficiency separations, with well resolved peaks.

Micelles are often compared to simple biological membranes which are now known to consist of stacked molecules with polar heads and hydrophobic tails. The information gained from chromatographic separations using micelles in the mobile phase may enable pharmacologists to understand, to a greater extent, the specific adsorption of drugs and other complex molecules across biological membranes.

Finally a better understanding of the effects of temperature, organic modifier and pressure on micelle stability is important if micellar chromatography is to develop and become an accepted method within the area of chromatography.

Micellar chromatography is an advance in methodology. The research outlined in this area hopefully adds to this methodology. As the famous botanist and chromatographer M. S. Tswett once said, "Every scientific advance is an advance in method".

Literature Cited

1.   Jonkman, J. H. G. Pharm. Weekbl., 1975, 110, 649.
2.   Higuchi, T., Michaelis, A. Anal. Chem., 1968, 40, 1925.
3.   Modin, R., Schill, G. Acta. Pharm. Suecica, 1967, 4, 301.
4.   Tomlinson, E., Jeffries, T. M., Riley, C. M. J. Chromatogr., 1978, 159, 315.
5.   Gloor, R., Johnson, E. L. J. Chromatogr. Sci., 1977, 15, 413.
6.   Wittmer, D. P., Nuessle, N. O., Haney, W. G. Anal. Chem., 1975, 47, 1422.
7.   Soad, S. P., Sartoni, L. E., Wittner, D. P., Haney, W. G. Anal. Chem., 1976, 48, 796.
8.   Knox, J. H., Jurand, J. J. Chromatogr., 1975, 110, 103.
9.   Hearn, M. T. W. Science, 1978, 200, 1168.
10.   Hearn, M. T. W. J. Chromatogr., 1978, 161, 291.
11.   Knox, J. H., Jurand, J. J. J. Chromatogr., 1976, 125, 89.
12.   Brown, N. D., Hall, L. L., Sleeman, H. K., Doctor, B. P., Demaree, G. E. J. Chromatogr., 1978, 150, 225.
13.   Moyer, T. P., Jiang, N. S. J. Chromatogr., 1978, 153, 365.

14.   Olieman, C., Maat, L., Waliszewki, K., Beyerman, H. C.
       J. Chromatogr., 1977, 133, 382.
15.   Knox, J. H., Laird, G. R. J. Chromatogr., 1976, 122, 17.
16.   Kraak, J. C., Jonker, K. M., Huber, J. F. K.
       J. Chromatogr., 1977, 49, 2295.
17.   Terwij-Groen, C. P., Heemstra, S., Kraak, J. C.
       J. Chromatogr., 1978, 161, 69.
18.   Horvath, C., Melander, W., Molnar, I., Molnar, P.
       Anal. Chem., 1977, 49, 2295.
19.   Hoffman, N. E., Liao, J. C. Anal. Chem., 1977, 49, 2231.
20.   Horvath, C., Melander, W., Molnar, I. J. Chromatogr.,
       1976, 125, 129.
21.   Van der Verne, J. L. M., Hendrikx, J. L. H. M.,
       Deedler, R. S. J. Chromatogr., 1978, 167, 1.
22.   Kissinger, P. T. Anal. Chem., 1977, 49, 883.
23.   Bidlingmeyer, B. A., Deming, S. N., Price, W. P.,
       Sachok, B., Petrusek, M. J. Chromatogr., 1979, 186, 419.
24.   Yarmchuk, P., Weinberger, R., Hirsch, R. F.,
       Cline Love, L. J. Anal. Chem., 1982, 54, 2233.
25.   Armstrong, D. W., Nome, F. Anal. Chem., 1981, 53, 1662.
26.   Dorsey, J. G., DeEchegaray, M. T., Landy, J. S.
       Anal. Chem., 1983, 55, 924.
27.   Yarmchuk, P., Weinberger, R., Hirsch, R. F.,
       Cline Love, L. J. J. Chromatogr., 1984, 283, 47.
28.   Small, H., Stevens, T. S., Bauman, W. C. Anal. Chem., 1975,
       47, 1801.
29.   Skelly, N. E. Anal. Chem., 1982, 54, 712.
30.   Iskandarani, Z., Pietrzyk, D. J. Anal. Chem., 1982,
       54, 1065.
31.   Molnar, I., Knauer, H., Wilk, D. J. Chromatogr., 1980,
       201, 225.
32.   Cassidy, R. M., Elchuk, S. J. Chromatogr., 1983, 262, 311.
33.   Cassidy, R. M., Elchuk, S. J. Chromatogr., 1983, 21, 454.
34.   Reeve, R. W. J. Chromatogr., 1979, 177, 393.
35.   de Kleijn, J. P. Analyst, 1982, 107, 223.
36.   Knox, J. H., Hartwick, R. A. J. Chromatogr., 1981, 204, 3.
37.   Hung, C. T., Taylor, R. B. J. Chromatogr., 1981, 209, 175.
38.   Mullins, F. G. P., Kirkbright, G. F. Analyst, 1984,
       109, 1217.
39.   Stiger, D. J. Phys. Chem., 1984, 68, 3603.
40.   Cruickshank, P. A., Jarrow, H. C. J. Agric. Food Chem.,
       1973, 21, 333.
41.   Vonk, J. W., Kaars Sijpestein, A. J. Envrion. Sci. Health,
       1976, B11, 33.
42.   Innes, J. R. M., Valerio, M., Ulland, B. M., Palotta, A. J.,
       Petrucelli, L., Fishbein, L., Hart, E. R., Falk. H. L.,
       Klein, M., Peters, A. J. J. Natl. Cancer Inst., 1969,
       42, 1101.
43.   Ulland, B. M., Weisberger, B. H., Weisberger, E. K.,
       Rice, J. M., Cypher, R. J. Natl. Cancer Inst., 1972,
       49, 483.
44.   Ottnad, M., Jenny, N. A., Roder, C. H. in Zweig, G. and
       Sherma,J. Editors, "Analytical Methods for Pesticides and
       Plant Growth Regulators", Volume 10, Academic Press,
       New York, 1978, p. 563.

45.   Smith, R. M., Moraji, R. L., Salt, W. G. Analyst., 1981, 106, 129.
46.   Smith, R. M., Moraji, R. L., Salt, W. G., Stretton, R. J. Analyst, 1980, 105, 184.
47.   Gustafsson, K. H., Thompson, R. A. J. Agric. Food. Chem., 1981, 29, 729.
48.   Gustafsson, K. H., Fahigren, C. H., J. Agric. Food Chem., 1983, 31, 463.
49.   Kirkbright, G. F., Mullins, F. G. P. Analyst, 1984, 109, 493.
50.   Mullins, F. G. P., Kirkbright, G. F. Analyst., 1986, in press.
51.   Mullins, F. G. P., Kirkbright, G. F. Analyst., 1986, in press.
52.   Almgren, M., Grieser, F. and Thomas, J. K. J. Am. Chem. Soc., 1979, 101, 279.
53.   Tagahira, S. Anal. Chem., 1983, 55, 730.
54.   Foley, J.P., Dorsey, J. G. Anal. Chem., 1983, 55, 730.

RECEIVED April 16, 1987

Chapter 5

Stationary Phase in Micellar Liquid Chromatography: Surfactant Adsorption and Interaction with Ionic Solutes

Alain Berthod, Ines Girard, and Colette Gonnet

Laboratoire des Sciences Analytiques, Universite Claude Bernard, Lyon 1, 69622 Villeurbanne cedex, France

The stationary phases play an important part in Liquid Chromatography using micellar mobile phases. They interact with both the surfactant and with solutes. To study the interactions with surfactants, adsorption isotherms were determined with two ionic surfactants on five stationary phases: an unbonded silica and four monomeric bonded ones. It seems that the surfactant adsorption closely approaches the bonded monolayer (4.5 μmol/m2) whatever the bonded stationary phase-polarity or that of the surfactant. The interaction of the stationary phase and solutes of various polarity has been studied by using the K_{SW} values of the Armstrong model. The K_{SW} value is the partition coefficient of a solute between the stationary phase and the aqueous phase. Methanol decreases the hydrophobic interactions, NaCl decreases the ionic interactions; so their influence on adsorption isotherms were compared with the modifications of the K_{SW} values. The retention of comicellizable and ionic solutes, used as surfactant tracer, has given information about the affinity of the surfactant for the stationary phases. The retention of ionic solutes has shown some ion-exchange capacity of the surfactant covered stationary phases. The retention of toluene has shown the role of the subjacent bonded moiety.

Micellar Liquid Chromatography (MLC) uses surfactant solutions as mobile phases for reversed phase liquid chromatography. The two main properties of surfactant molecules, as related to chromatography, are micelle formation and adsorption at interfaces. The micelles play the role of the organic modifier, so their influence on retention has been extensively studied ($\underline{1}$). At surfactant concentrations above the critical micellar concentration (CMC), micelles are present and the amount of free surfactant is essentially

0097-6156/87/0342-0130$06.00/0
© 1987 American Chemical Society

constant and equal to CMC. It has been assumed and demonstrated that the amount of surfactant adsorbed on the stationary phase is constant at concentrations above the CMC (2-3).
The aim of the present work was to study the adsorption of two ionic surfactants on five stationary phases of various polarities in order to elucidate the role of the stationary phase in the retention mechanism of MLC. The effect of two additives, methanol and sodium chloride, has also been investigated.

Experimental Section

Surfactants. The two ionic surfactants were sodium dodecylsulfate (SDS) and cetyltrimethylammonium bromide (CTAB). Their physico-chemical properties were reported in Table I.

Table I. Physicochemical properties of the studied surfactants.
N=aggregation number, β =counterion binding

Surfactant	medium	CMC (mol/l)	N	β
SDS	water	8.2×10^{-3}	62	0.65
mw=288.4	water+methanol 95%-5% v/v	8.0×10^{-3}	~60	0.6
V=0.246 L/mol	water+NaCl 0.1 mol/L	1.4×10^{-3}	~80	0.95
CTAB	water	8×10^{-4}	90	0.84
mw=364.5	water+methanol 95%-5% v/v	9×10^{-4}	~70	<0.84
V=0.364 L/mol	water+NaCl 0.1 mol/L	2×10^{-4}	>90	0.90

The effect of methanol on micellar solutions is slight at the low concentration used (5% v/v = 1.3 mol/L = 0.022 mole fraction). The effect of NaCl however, is more significant: the CMC is greatly decreased, the degree of counterion binding and the aggregation number are increased.

Stationary phases. Five stationary phases from Shandon (Runcorn, Cheslvie, GB) were used; they were spherical microparticules of 5 µm mean diameter. The four bonded silicas were manufactured from the same parent silica (Hypersil) and possess a monolayer coverage of trimethylsilyl (SAS Hypersil), dimethyloctylsilyl (MOS Hypersil), octadecylsilyl (ODS Hypersil) and cyanopropylsilyl (CPS Hypersil) groups. Their physicochemical properties are listed in Table II. The elemental analysis of carbon (%C), corrected for the %C value of naked silica, enables to estimate the surface concentration of the substituent (Γ) with:

$$\Gamma = (10^6 \%C/S)/(1200*n_C - \%C*(M-1))$$

in which S is the specific surface area of the **parent** silica and n_C and M are, respectively, the carbon number and the molecular weight of the bonded moiety. For CPS and ODS Hypersil, the bonding reagent was not well-known and information from Hypersil supplier (Shandon) was incomplete. So, for these two stationary phases, the calculation of Γ has been performed assuming pure cyanopropyldimethylsilyl and pure octadecyldimethylsilyl bonded moiety. The Γ values of Table II are only indicative. Anyway, the calculated values of Γ closely approach the highest limiting concentration of a bonded monolayer (4.5 μmol/m^2).

Solutes. Toluene, although polarizable, was chosen as an apolar solute. Caffeine was chosen as a polar but nonionic solute. Four ionic solutes were tested: benzyltrimethylammonium bromide (BTAB) is a cationic quaternary ammonium salt. Benzoic acid acts as an anionic solute at mobile phase pH values between 5.5 and 6.5 (the pK_A lies between 3.7 in CTAB solutions and 4.7 in SDS solutions)(4). Sodium paraoctylbenzene sulfonate (SOBS) (pK_A=0.8) and cetylpyridinium chloride (CPC) were chosen as ionic solutes having surfactant properties. Their hydrophobic "tails" have the same lenghts as those of SDS and CTAB, respectively.

Table II. Physicochemical properties of the studied silica.
S=specific surface area, %C=carbon percentage,
Γ =surface coverage

Trade name	Bonded moiety	S m^2/g	%C w/w	Γ μmol/m^2
Hypersil	Unbonded	150	0.3	-
CPS Hypersil	Cyanopropyl	115	4.2	(4.5)
SAS Hypersil	Trimethyl	104	18	4.5
MOS Hypersil	Octyl	129	24	4.1
ODS Hypersil	Octadecyl	105	24	(2.1)

Results and discussion

Adsorption isotherms. **Pure aqueous mobile phases:** The surfactant adsorption on the stationary phase could occur in at least two ways (5): i-Hydrophobic adsorption; the alkyl tail is adsorbed and the ionic head group would then be in contact with the polar solution, ii-Silanophilic adsorption; the ionic head group is adsorbed and the stationary phase becomes more hydrophobic (Figure 1).

With the exception of SDS on naked silica, all the curves are of the H type (6); i.e. the amount of adsorbed surfactant increases rapidly and reaches a plateau for surfactant concentrations higher than the CMC. Two remarks should be made here: the first one is that the adsorption plateaus are, unexpectedly, very close to each other for C1, C8 and C18 bonded phases. The second remark is that the maximum adsorption is obtained on SAS (C1) Hypersil but not on the more hydrophobic ODS (C18) phase (Table III).

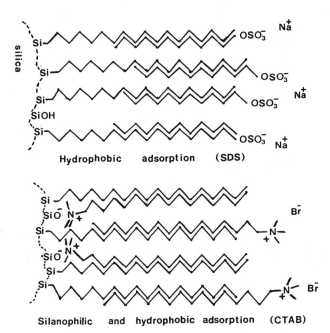

Figure 1:   The two possible ways for surfactant adsorption onto ODS Hypersil (C18 monomer bonded silica).

    For SDS adsorption, a slight slope exists (except on ODS
silica), that shows a further adsorption of SDS in the presence of
very high micellar concentration; such a comportment has been noted
by Hinze (7) with Brij 35, a nonionic surfactant.

Table III. Ionic surfactant adsorption in $\mu mol/m^2$ on five
stationary phases. Temperature $25^\circ C$, experimental
error 5%

Surfactant	Concentration mol/L	silica	CN	C1	C8	C18
	0.1	0	2.1	4.6	3.7	4.7
SDS	0.3	0.5	3.0	5.2	4.2	4.9
	0.4	1.5	3.2	5.4	4.4	5.0
	0.05	2.0	3.4	4.6	3.8	4.6
CTAB	0.1	2.0	3.6	4.8	3.8	4.6
	0.2	2.0	4.0	5.0	3.8	4.7

Additive effects: Low amounts of methanol have little effect on the
free surfactant concentration (Table I). Nevertheless, the adsorbed
amount of surfactant is decreased by 5% v/v methanol. The decreasing
is slight (8%) with ODS Hypersil and the two surfactants and with
SAS Hypersil and CTAB, but it can reach 30% in the case of SDS on
SAS Hypersil (Table IV).

Table IV. Effect of 5% v/v methanol and 0.1 mol/L NaCl on the
ionic surfactant adsorption in $umol/m^2$ on SAS (C1)
and ODS (C18) Hypersil stationary phases at $25^\circ C$.
Experimental error 5%

Surfactant	Conc. mol/L	SAS Hypersil (C1) water	methanol	NaCl	ODS Hypersil (C18) water	methanol	NaCl
	0.1	4.6	3.8	4.2	4.7	4.4	5.2
SDS	0.3	5.2	4.5	4.7	4.9	4.5	5.3
	0.4	5.4	5.0	4.9	5.0	4.6	5.4
	0.02	4.5	4.2	4.7	4.6	4.3	4.6
CTAB	0.05	4.6	4.4	5.2	4.6	4.3	4.6
	0.1	4.8	4.7	5.5	4.6	4.3	4.6

These differences can be rationalized by taking into account the
physicochemical structure of the bonded ODS layer. According to
Scott and Simpson (8), the "collapsed state" of the ODS layer was
destroyed when about 5% v/v methanol was present in the mobile

phase. When the alkyl chains of the ODS bonded layer return again to the brush-form, the specific surface is increased and/or silanols becomes accessible. Methanol reduces the hydrophobic interactions and decreases the amount of surfactant adsorbed. In the case of ODS Hypersil, this decrease is partially compensated by the disappearance of the "collapsed state". The setting upright of the alkyl-chains allows the insertion of surfactant molecules.

Assuming that the adsorbed amount of surfactant is only dependent upon the free surfactant concentration, as the added NaCl decreases the CMC (Table I), it was expected to find an adsorbed amount of surfactant lower with NaCl than with pure aqueous mobile phase. However, that was not observed: the CMC values are about five times lower with NaCl and the adsorbed amount of surfactant was only 10% lower, for SDS on SAS (C1) Hypersil, or equal, for CTAB on ODS (C18) Hypersil, and even 10% **higher** for CTAB and SDS on SAS and ODS Hypersil, respectively. This effect has been studied in detail by Bartha et al. ($\underline{9}$). They have shown a linear increase of adsorbed amount of sodium butanesulfonate on ODS Hypersil, at constant mobile phase concentration of the pairing ion, with increasing sodium concentration. This "salting out" effect lowers the ionic repulsions and enhances the hydrophobic interactions. The "salting out" effect is greater on SDS because the common ion effect. This may be the reason why SDS adsorption on ODS Hypersil is higher with NaCl than without, and the CTAB adsorption is equal on ODS with NaCl or without. On SAS Hypersil, in addition to the "salting out" effect, there is a possible ion-exchange phenomena with the accessible surface-silanols. These silanols have much greater affinity for CTA^+ than for Na^+ ($\underline{11}$) and no affinity for anionic DS^-. This produces an amount of adsorbed CTAB, with NaCl, greater than the one with pure water (Table IV) and the reverse for SDS adsorption on SAS silica.

Retention study. At surfactant concentrations below CMC, micelles do not exist and, as demonstrated by Knox ($\underline{12}$), Deming ($\underline{13}$) and our previous works ($\underline{14-15}$), the degree of retention was directly related to the surface charge arising from the adsorbed surfactant . With both the surfactants, the retention of neutral species (toluene and caffeine) slightly decreased. When an anionic surfactant was adsorbed, the retention of negatively charged solutes (benzoate and SOBS) fell dramatically whereas the retention of cationic solutes (BTAB and CPC) increased. The reverse occured with the cationic surfactant ($\underline{14}$). The same kind of behavior was observed with pure aqueous mobile phases, 5-95% v/v methanol-water phases and 0.1 mol/L NaCl phases.

Armstrong ($\underline{16}$) proposed a classification of the solutes according to their chromatographic properties in the micellar mobile phases: solutes binding to the micelles, nonbinding solutes and anti binding solutes. On the studied stationary phases, benzoic acid (at pH values of 6) behaved as a nonbinding solute or as an antibinding solute in SDS systems, i.e. its retention was constant or slightly increased, but it is difficult to measure accurately due to the very low k' values (less than 0.3). Benzoic acid was a highly binding solute with CTAB mobile phases. BTAB was a nonbinding solute in the presence of CTAB micellar mobile phases and a highly binding solute in SDS micellar mobile phases. The electrostatic repulsions may be responsible for the nonbinding character, but the binding character can occur in spite of electrostatic repulsions. Indeed, the negative SOBS solute behaved as a binding solute with anionic SDS mobile phases, and the positive CPC solute behaved as a binding solute in cationic CTAB mobile phases. SOBS and CPC are binding solutes by comicellization. They form mixed micelles with SDS and CTAB, respectively.

The equation derived for the retention of binding solutes was:

$$\frac{1}{k^-} = \frac{1}{\Phi} \left[\frac{V \, (K_{MW} - 1)}{K_{SW}} C_m + \frac{1}{K_{SW}} \right]$$

in which k^- is the capacity factor of the studied solute,
    V is the molar volume of the surfactant (Table I),
    Φ is the phase ratio V_s/V_o,
    V_s is the stationary phase volume,
    V_o is the void volume,
    C_m is the concentration of surfactant in the micellar form
        (i.e. the total surfactant concentration minus the CMC),
K_{MW} and K_{SW} are the dimensionless solute partition coefficients
        between micelles and the bulk water and between the
        stationary phase and the bulk water, respectively.
The plot of $1/k^-$ versus the micellar concentration gave straight lines whose slopes and intercepts allow to calculate the K_{MW} and K_{SW} values to be obtained. K_{MW} values measure the solute affinity for micelles. K_{MW} should be independent of the nature of the stationary phase in the same mobile phase. K_{SW} values give information about the affinity of the solute for the surfactant covered stationary phase.

K_{MW} values. As expected, the K_{MW} values were almost independent of the nature of the stationary phase in the same micellar mobile phase. Table V presents the mean K_{MW} values obtained with five sets of chromatograms using the same mobile phase and the five different stationary phases described (noted SDS and CTAB) or the mean K_{MW} values obtained with two sets of chromatograms using the same mobile phase and only the two stationary phases :SAS and ODS Hypersil (mobile phases with methanol and NaCl additives).
The Armstrong model (17) can describe the retention of apolar, polar and even ionic solutes, provided they were binding solutes. The highest K_{MW} values corresponded to electrostatic interactions (1600 for CTAB with SDS micelles and 2600 for benzoic acid with CTAB with SDS micelles and 2600 for benzoic acid with CTAB micelles) or to comicellization (190 for SOBS with SDS micelles and 3000 for CPC with CTAB micelles).
Methanol decreases the K_{MW} values of all the binding solutes because it increases the hydrophobic interactions. The effect of NaCl is very different; the influence on the K_{MW} values of nonionic solutes was slight (Table V) and the influence on the K_{MW} values of ionic solutes was important. The K_{MW} values of BTAB with anionic micelles and the one of benzoic acid with cationic micelles are significantly decreased. As NaCl decreases the electrostatic interactions, this could be interpreted as an evidence of the ionic binding of these two solutes towards micelles. The K_{MW} values of the comicellizable solutes were increased by NaCl. The SOBS values

is three times higher with NaCl because of the Na^+ common ion effect. The CPC value is only 20% higher in CTAB solution with NaCl 0.1 mol/L. NaCl promotes the micelle-formation.

Table V. K_{MW} values. Experimental error 30%

Medium	Solutes	NONIONIC SOLUTES		IONIC INTERACTION	COMICELLIZABLE SOLUTES
		Toluene	Caffeine	BTAB	SOBS
SDS		240	40	1600	190
SDS + methanol 5%		200	18	880	150
SDS + NaCl 0.1 M		280	45	540	530
		Toluene	Caffeine	Benzoic ac.	CPC
CTAB		300	35	2600	3000
CTAB + methanol 5%		220	20	2100	2200
CTAB + NaCl 0.1 M		320	37	610	3500

K_{SW} values. K_{SW} values give information about the affinity of the solute for the surfactant covered stationary phases. The first observation indicates that, in spite of this surfactant-coverage, the polar nature of the bonded stationary phases is preserved: the order of increasing K_{SW} values of toluene is the same order as the decreasing stationary phase-polarity i.e. silica $\ll$ CN $\sim$ C1 $\ll$ C8 $\sim$ C18. The K_{SW} values of toluene, with SDS mobile phases, were slighty lower than those with CTAB mobile phases (Table VI). Toluene, an apolar solute, seems to have a higher affinity for the CTAB covered phases than for the SDS covered phases. If we consider the two processes of surfactant adsorption (Figure 1), the silanophilic process is important in CTAB adsorption given the great affinity of quaternary ammonium for surface-silanols (11). As a result, the stationary phase becomes more hydrophobic with CTAB than with SDS, which could explain the magnitude of the K_{SW} values for toluene. Caffeine is a polar solute, its K_{SW} values on the more hydrophobic stationary phases (MOS and ODS Hypersil) were weak and much lower than the respective toluene-K_{SW} values. The caffeine K_{SW} values were greater on SDS covered stationary phases than on CTAB covered ones. The explanation in the case of toluene holds true, that is to say caffeine has more affinity for the more polar SDS covered stationary phases (9.6 on C8, for example) than for the same phase but CTAB covered (1.8 on C8 with CTAB) (Table VI).

The effect of the two additives is more evident with the ionic solutes. Nonetheless, the effect of methanol was a slight decrease in all the K_{SW} values. NaCl increases the hydrophobic interactions of toluene with the stationary phases.

Table VI. K_{SW} values of the nonionic solutes.
Experimental error 20%

Solute	Mobile phase	Silica	CN	C1	C8	C18
			Stationary	phases		
TOLUENE	SDS	1.3	23	19	150	140
	SDS+methanol 5%			13		125
	SDS+ NaCl 0.1 M			46		155
	CTAB	0.35	63	55	190	190
	CTAB+methanol			50		125
	CTAB+ NaCl 0.1M			90		205
CAFFEINE	SDS	1.0	3.5	12	9.6	5.6
	SDS+methanol 5%			3.1		4.1
	SDS+ NaCl 0.1 M			11		7.8
	CTAB	0.28	2.0	3.6	1.8	2.3
	CTAB+methanol			2.6		1.3
	CTAB+ NaCl 0.1M			3.6		1.9

Table VII. K_{SW} values of the ionic solutes.
Experimental error 20%

Solute	Mobile phase	CN	C1	C8	C18
			Stationary	phases	
IONIC INTERACTIONS					
BTAB	SDS	51	420	1400	1500
	SDS+methanol 5%		370		1800
	SDS+ NaCl 0.1 M		140		840
	CTAB		non binding solute		
Benzoic acid	SDS		non binding solute		
	CTAB	490	870	760	550
	CTAB+methanol 5%		840		460
	CTAB+ NaCl 0.1 M		190		180
COMICELLIZATION					
SOBS	SDS	10	9.4	35	33
	SDS+methanol 5%		7		21
	SDS+ NaCl 0.1 M		45		77
	CTAB		too much solute retained		
CPC	SDS		too much solute retained		
	CTAB	910	850	1000	610
	CTAB+methanol 5%		460		460
	CTAB+ NaCl 0.1 M		340		930

In the case of ionic solutes, the stationary phase exhibited some ion-exchange capacity. In micellar mobile phases of the opposite charge, the comicellizable solutes were so much retained that no peaks were obtained after three days of elution at 1 mL/mn. BTAB and benzoic acid acted like nonbinding solutes in the micellar mobile phase of the same charge (CTAB and SDS, respectively); their k' values were lower than 0.3 (Table VII).

The K_{SW} values of BTAB in SDS mobile phases were similar to the one of toluene. It seems that BTAB was retained as an apolar ion-pair. The order of increasing K_{SW} values of BTAB corresponds to the decreasing stationary phase-polarity. Benzoic acid in CTAB mobile phases seems to be also retained as an ion-pair. 5% methanol slightly decreased the K_{SW} values. The effect of NaCl was much more significant: the decrease was 50% (BTAB with SDS mobile phases) up to 80% (benzoic acid with CTAB mobile phases). Sodium and/or chloride ions decrease the ionic interactions and the ion-exchange capacity.

The K_{SW} values of the comicellizable solutes seem to show the great affinity of quaternary ammonium for surface-silanols. The CPC values were two orders of magnitude higher than the SOBS K_{SW} values. Methanol decreased these K_{SW} values from about 30%. NaCl increased the SOBS K_{SW} values from two times on ODS Hypersil up to five times on SAS Hypersil ("salting out" with common ion effect). The presence of NaCl increased the CPC K_{SW} values on ODS Hypersil from about 50%, but it decreased the same value on SAS Hypersil from 60%. The surface silanols of SAS Hypersil are much more accessible than the ones of ODS Hypersil. Sodium ions compete with CPC on SAS Hypersil-silanol groups, that could explain the NaCl effect on the K_{SW} values of CPC (Table VII).

It must be noted that the K_{SW} values of comicellizable solutes were not clearly related to the global amount of surfactant adsorbed.

Effect on retention time. For a given solute, if the K_{MW} value increases, the retention time of the solute decreases. This can be interpreted as resulting from micelle-transport. If the K_{SW} value increases, the solute affinity for the stationary phase and the retention time increases. For all binding solutes, the retention time decreases as the micellar concentration increases. The effect of the two additives studied has given unusual retention-behavior. For example, caffeine, in SDS micellar mobile phases, illustrated this unusual behavior. On ODS Hypersil, the retention of caffeine was **decreased** by methanol addition at low micellar concentration because the K_{SW} was decreased, but the retention was **increased** by methanol addition at high micellar concentration because the micelle-transport is more important in pure aqueous solutions than in methanolic solutions (Figure 2). The same kind of behavior is observed with ionic benzoic acid in CTAB mobile phases towards methanol or NaCl addition.

Conclusion

In the studied cases: monomeric bonded silica and ionic surfactants, the surfactant adsorption was almost constant with differing

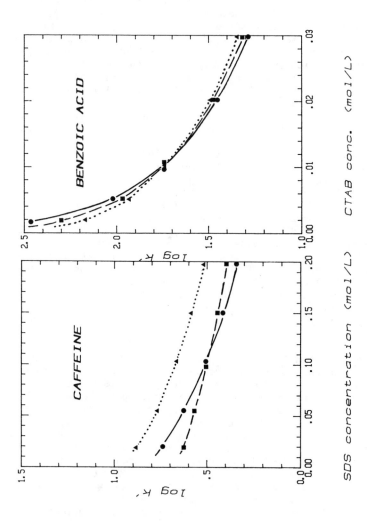

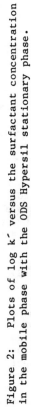

Figure 2:   Plots of log k´ versus the surfactant concentration
in the mobile phase with the ODS Hypersil stationary phase.
        solid   lines: aqueous mobile phases.
        dotted lines: NaCl 0.1 mol/L mobile phases.
        dashed lines: methanol-water 5-95 % v/v.

surfactant concentration above CMC, with, in some cases, an increase of less than 20%. Because of this, the model described in the literature (17) fits not only the retention of polar and apolar solutes but also the retention of ionic solutes provided they bind to micelles either by electrostatic interactions or by comicellization. The total amount of surfactant adsorbed on the bonded silicas closely approaches the highest limiting concentration of a monolayer given to be 4.5 $\mu mol/m^2$ by Kovats (18). That is why the amount of anionic or cationic surfactant adsorbed on SAS Hypersil (C1) or on ODS Hypersil (C18) was found to be about 4.6 $\mu mol/m^2$ although the polarity of SAS Hypersil was higher than that of ODS Hypersil and although the properties of the two surfactants were very different.

Acknowledgments

This work was supported by the Centre National de la Recherche Scientifique (C.N.R.S.), unité associé 07 0435 (J.M. Mermet).

Literature Cited

1. Armstrong, D.W. Sep. Purif. Methods 1985, 14, 213-304.
2. Yarmchuck, P.; Weinberger, R.; Hirsh, R.F.; Cline Love, L.J. Anal. Chem. 1982, 54, 2233-2238.
3. Dorsey, J.G.; Khaledy, M.G.; Landy, J.S.; Lin, J.L. J. Chromatogr. 1984, 316, 183-191.
4. Arunyanart, M.; Cline Love, L.J. Anal. Chem. 1985, 57, 2837-2843.
5. Nahum, A.; Horvath, Cs. J. Chromatogr. 1981, 203, 53-63.
6. Giles, C.H. In "Anionic Surfactants", Lucassen-Reynders, E.H. Ed., Marcel Dekker, N.Y., 1981, Vol. 11, Chapter 4.
7. Borgerding, M.F.; Hinze, W.L. Anal. Chem. 1985, 57, 2183-90.
8. Scott, R.P.; Simpson, C.F. J. Chromatogr. 1980, 197, 11-20.
9. Bartha, A.; Billiet, H.A.H.; De Galan, L.; Vight, G. J. Chromatogr. 1984, 291, 91-102.
10. Tramposch, W.G.; Weber, S.G. Anal. Chem. 1984, 56, 2567-2571.
11. Papp, E.; Vight, G. J. Chromatogr. 1983, 282, 59-70.
12. Knox, J.H.; Hartwick, R.A. J. Chromatogr. 1981, 204, 3-21.
13. Tang, M.; Deming, S.N. Anal. Chem. 1983, 55, 425-428.
14. Berthod, A.; Girard, I.; Gonnet, C. Anal. Chem. 1986, 58, 1356-1358.
15. Berthod, A.; Girard, I.; Gonnet, C. Anal. Chem. 1986, 58, 1362-1367.
16. Armstrong, D.W.; Stine, G.Y. Anal. Chem. 1983, 55, 2317-2320.
17. Armstrong, D.W.; Nome, F. Anal. Chem. 1981, 53, 1662-1666.
18. Gobet, F.; Kovats, E. Sz. Adsorp. Sci. Technol. 1984, 9, 77-89.

RECEIVED October 24, 1986

Chapter 6

Micellar Electrokinetic Capillary Chromatography

M. J. Sepaniak[1], D. E. Burton[1], and M. P. Maskarinec[2]

[1]Department of Chemistry, University of Tennessee, Knoxville, TN 37996-1600
[2]Analytical Chemistry Division, Oak Ridge National Laboratory, P.O. Box X, Oak Ridge, TN 37830

The incorporation of micelles in the mobile phase in capillary zone electroporesis permits the efficient separation of a variety of neutral compounds. Efficiencies in excess of 100,000 plates/m are routinely attained. The mass transport processes which are important in micellar electrokinetic capillary chromatography are described, along with the technique. The technique is particularly useful for biological separations. Preliminary data and discussion related to column selectivity and efficiency are presented.

Micellar electrokinetic capillary chromatography, MECC, was first reported by Terabe, et.al. (1). The technique combines many of the operational principles and advantages of micellar liquid chromatography (2) and capillary zone electrophoresis, CZE (3). CZE is performed using narrow-bore capillary columns (CA. 50 μm i.d. x 100 cm) which are filled with an aqueous buffer solution. A large applied electric field drives charged sample solutes, which are injected as a sharp plug, toward the detection end of the column. Efficiencies in excess of 100,000 plates/m are generally observed. Electroosmotic flow (4) provides another means of transporting solutes, including neutrals, through the column.

In CZE, differences in the viscous drag of neutral solutes, primarily as a result of size differences, can provide for their separation (3). However, these differences are usually very small and, consequently, the technique is not very useful for separating neutral compounds. With the MECC technique, a surfactant is added to the mobile phase at a concentration above its critical micelle concentration. The resulting micelles provide an effective mechanism for separating neutral compounds. Neutral solutes are separated based on their differential partitioning between an electroosmotically-pumped mobile phase and the hydrophobic interior of the

0097-6156/87/0342-0142$06.00/0
© 1987 American Chemical Society

micelles, which are charged and moving at a velocity different from
that of the mobile phase due to electrophoretic effects.

The mass transport phenomena operative in MECC are depicted in
the expanded view of a capillary section shown in Figure 1. The
column contains the surfactant/buffer solution. Negatively charged
(anionic) surfactant is considered in the figure. Primary and
secondary sorbed ions generate an electric double layer potential
(zeta potential) at the inside surface of the capillary. The
larger the zeta potential the greater the disparity in the overall
mobilities of positive and negative ions. Because of this disparity a net flow of solvent (electroosmotic flow) results when an
electric field is applied and solvated ions move toward the electrode of opposite sign. In our work with both anionic (e.g.,
sodium dodecyl sulfate, SDS), and cationic (e.g., cetyltrimethylammonium chloride, CTAC) surfactants, the electroosmotic flow velocity (Veo) opposes the electrophoretic velocity of the micelles
(Vm,e) and is of a greater magnitude. Consequently, two distinct
phases, the mobile phase and the micellar phase, exist within the
column and migrate at different velocities toward the electrode
with the same charge as the micelles. Veo is proportional to the
magnitude of the zeta potential and the applied potential, while
Vm,e is proportional to the micellar electrophoretic mobility and
the applied potential. A solute (S) which partitions between the
mobile and micellar phases will have a band velocity (Vs) that is
intermediate between Vm and Veo.

Chromatograms in MECC differ from those observed in conventional elution chromatography in that there is generally a limited
elution range. Retention times, t_R, in MECC are given by equation 1 where t_o is the retention time of a solute which is not

$$t_R = \frac{t_o}{R + (t_o/t_m)(1-R)} \tag{1}$$

solubilized by the micelles, t_m is the retention time of a solute
which is completely solubilized, and R is the fraction of solute
not solubilized (i.e., the MECC analog of the retention ratio in
conventional LC). The limited elution range of MECC translates
into relatively short separation times, but adversely, limits the
peak capacity of the technique.

Certain fundamental characteristics of MECC that influence
retention have been investigated (5). The technique has been used
in the analysis of a variety of samples including phenolic compounds (1), phenylthiohydantoin-amino acids (6), and metabolites of
vitamin B_6 (7). In related electrokinetic separation techniques,
substituted benzene compounds have been separated based on the
formation of inclusion complexes with an ionic cyclodextrin derivative in the mobile phase (8) and polyaromatic hydrocarbons have
been separated based on solvophobic interactions with a tetraakyl-
ammonium ion in the mobile phase (9). The effects of injection procedures on efficiency have also been studied (10).

Experimental

Apparatus. The experimental configuration used in this work is shown in Figure 2. Columns were 25-100 μm i.d. fused silica capillaries which were protected by a polimide coating. The column length was generally 50 - 100 cm. A 1 - 2 mm section of the protective coating was removed near one end of the column to provide an optical window for laser-based fluorescence detection (10) using the 488 nm line of a Cyonics Model 2001 Ar$^+$ laser for excitation, or UV absorbance detection using a Jasco UVIDEC-100-III detector with a modified flow cell compartment (10). High voltage (0 - 40 kV) was supplied by Hipotronics regulated DC power supplies (Models R30B and R40B). Electrodes were either platinum wire or graphite rods.

Procedures. The capillary columns were rinsed with 0.1 M HCl prior to filling with mobile phase. Typical mobile phases were 0.01 M SDS and 0.01 M Na$_2$HPO$_4$, in the case of negatively charged micelles, and 0.02 M CTAC, 0.01 M Na$_2$HPO$_4$, and 0.01 M Na$_2$B$_4$O$_7$, in the case of positively charged micelles. However, the surfactant concentrations were varied in a column efficiency study and buffer concentrations were often adjusted to keep currents at levels that minimize heating.

Electroinjection was used to introduce sample into the column (10). With this technique the surfactant/buffer reservoir at the inlet of the column is replaced by the sample solution. High voltage (2 - 5 kV) is applied for 5 - 30 seconds. The surfactant/buffer reservoir is then returned and the high voltage reapplied to effect the separation.

Chemicals. The surfactants used in this work were obtained from Sigma Chemical Co. and mobile phases were prepared using distilled and deionized water. The purines were obtained from B-L Biochemicals and Merck Chemicals. The buffers and the other test solutes were obtained from Fisher Scientific. The amines were derivatized with 7-chloro-4-nitrobenzo-2,1,3-oxadizaole (NBD-Cl) from Regis Chemicals using a procedure supplied by the manufacturer.

Results and Discussion

MECC separations are generally limited to compounds which are reasonably soluble in the mobile phase. In the case of normal micelles the sample components to be separated must have some solubility in the aqueous phase. It is interesting to note that good sample solubility in the aqueous-micelle mixture does not assure an effective separation. The addition of micelles to an aqueous solution can greatly increase the solubility of hydrophobic compounds. For example, the solubility of pyrene in water is enhanced by 10^5 when the water is made 0.07 M in SDS (2). However, all of the hydrophobic compounds in a mixture tend to be nearly completely solubilized by the micelles and elute from a MECC column poorly resolved, with retention times near t_m.

Nevertheless, we have utilized the MECC technique for efficient separations of a variety of samples. For example, Figure 3 is the chromatogram of a separation of a mixture of benzene compounds substituted with groups which differ greatly in their

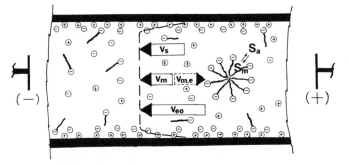

Figure 1. Expanded view of a MECC capillary section showing column dynamics.

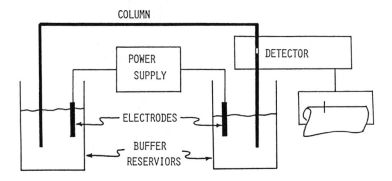

Figure 2. Diagram of apparatus.

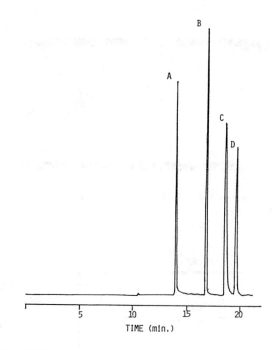

Figure 3. MECC chromatogram of aniline(A), nitrobenzene(B), phenol(C), and toluene(D).

chemical properties. The separation was performed with a 60 μm i.d. x 100 cm column containing 0.05 M CTAC, 0.01 M Na_2HPO_4, and 6 x 10^{-3} M $Na_2B_4O_7$. UV detection at 220 nm was employed. As expected the most hydrophobic component, toluene is eluted last. The weakly basic aniline is eluted first using this cationic micelle system, while the weakly acidic phenol is well-retained and somewhat broadened relative to the other components, both effects presumably due to electrostatic interactions with the positively charged micelles.

Biological samples are particularly amenable to MECC separation (7). Figure 4 is the chromatogram of a separation of a mixture of purines on a 75 μm i.d. x 100 cm column containing 0.01 M SDS and 0.01 M Na_2HPO_4. Detection was at 280 nm using the absorbance detector. The purines are efficiently separated in less than 20 minute using an applied potential of 20 kV. A higher potential could be employed to reduce the separation time but some loss in efficiency would occur (see discussion below).

The object in any separation is the resolution of the components of a sample. Resolution, R_s, in MECC is given by Equation 2 (5) where N is the number of theoretical plates for the

$$Rs = \frac{N^{1/2}}{4} (\frac{\alpha - 1}{\alpha}) (\frac{k'}{1 + k'}) (\frac{1 - t_o/t_m}{1 + (t_o/t_m)k'}) \qquad (2)$$

column, k' is the capacity factor (i.e., moles of solute in micellar phase divided by moles in the mobile phase), and α is the selectivity factor (k'_2/k'_1) for adjacent peaks in a chromatogram. Resolution in MECC depends on four factors, efficiency (N), selectivity (α), capacity (k'), and elution range (as reflected in t_o/t_m). The last factor illustrates an important difference between MECC and conventional elution chromatography which employs a true stationary phase. The optimum k' in MECC depends on t_o/t_m but is generally in the range of 2 - 5. As t_o/t_m approaches zero the last term in Equation 2 drops out and the MECC technique resembles conventional elution chromatography.

We are currently studying experimental factors which influence selectivity and elution range. The importance of these factors is demonstrated in Figure 5 which is the chromatogram of several nitrated polyaromatic hydrocarbons that were separated using the same conditions as in Figure 4 except that the detection wavelength was 230 nm. Chromatographic efficiency for this separation is excellent but retention selectivity is lacking. The same solutes can easily be resolved by conventional reversed phase LC, even though that technique exhibits much lower column efficiecy. The poor selectivity using MECC for this sample could be a general result of the large k' values for the solutes, or it could indicate poor retention discrimination among the solutes using this particular anionic micellar system.

As with other forms of chromatography, maximizing column efficiency is critical to the overall development of the technique. The majority of our research in MECC has focused on studies of experimental factors which influence column efficiency (11).

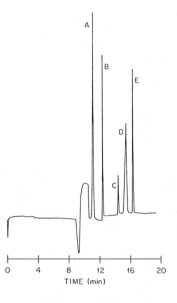

Figure 4. MECC chromatogram of the purines adenine(A),
6-methylpurine(B), 2 hydroxypurine(C), 6,6-dimethylamino-
purine(D), and xanthine(E).

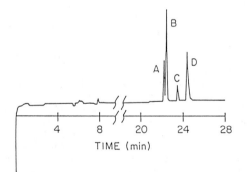

Figure 5. MECC chromatogram of 1,4-dinitronaphthalene(A),
1-nitronaphthalene(B), 9-nitroanthracene(C), and 1-nitro-
pyrene(D).

The "plug-like" velocity flow profile for electrokinetically pumped capillary columns (see Figure 1) is important in minimizing resistance to mass transfer within the mobile phase (4). Hydrostatically-pumped capillaries, have parabolic flow profiles which tend to severely disperse solute bands unless extreme narrow-bore (i.d.s less than 10 μm) capillaries are employed (12). Fortunately, larger capillaries, with less stringent detector volume requirements, can be efficiently used in MECC.

Under the proper conditions, column efficiency in MECC is outstanding. In one separation of purine compounds we obtained over 600,000 plates/m for theophylline. However, the effects of parameters such as column dimensions, applied voltage, and the concentration of the buffer and surfactant on efficiency can be very dramatic. A brief discussion of how these parameters influence efficiency follows.

Narrow-bore capillary tubes dissipate the heat generated in the electrophoretic process very efficiently. Nevertheless, we have observed heating effects when power dissipations exceed about 2 watts/m. This occurs for 75 μm i.d. x 100 cm columns, containing 10^{-2} M buffer, when applied voltages exceed about 30 kV. The heat generated produces a transverse temperature gradient within the column. Since electrophoretic mobility increases with temperature the "plug-like" flow profile shown in Figure 1 is distorted and column efficiency is degraded.

MECC separations are conducted in open capillaries, hence eddy diffusion is not problematic. However, the columns behave in many ways like packed columns, with the micelles functioning as uniformly sized and evenly dispersed packing particles. In packed columns, resistance to mass transfer in the mobile phase is reduced (i.e., efficiency improved) when smaller particles are used because the "diffusion distance" between particles is decreased. Average "inter-micellar" distance is the analogous parameter in MECC. This distance can be decreased by increasing surfactant concentration. In preliminary experiments, increasing SDS concentrations from just above its critical micelle concentration (about 7×10^{-3} M) to 5×10^{-2} M resulted in 5 - 10 fold decreases in plate heights for test solutes.

Despite the improved mass transfer characteristics of the "plug-like" flow profiles observed in MECC, "intra-column" resistance to mass transfer is significant at higher flow velocities (i.e., at high applied voltages). Although not as dramatic as in our work with hydrostatically-pumped open capillary LC, we have observed improvements in efficiency with the MECC technique when column diameter is reduced. This is illustrated in Figure 6. Peaks A and B correspond to the NBD-Cl derivatives of ethylamine and cyclohexylamine separated on a 75 μm i.d. column and detected by laser fluorometry. Peak C is NBD-cyclohexylamine from a 25 μm i.d. column of the same length. Efficiency is about a factor of 4 better with the smaller diameter column.

The applied voltage used to obtain Figure 6 was 20 kV. The effect is even more dramatic at higher voltages which result in greater flow rates. Optimum voltages for the columns used in this

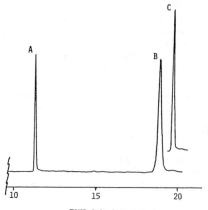

TIME (min.)

Figure 6. MECC chromatogram of NBD-ethylamine(A) and
NBD-cyclohexylamine(B), using a 75 μm i.d. column and
NBD-cyclohexylamine using a 25 μm i.d. column. The
mobile phase contained 0.01 M SDS and 0.003 M Na_2HPO_4.

work were about 10 kV, which corresponded to an electroosmotic flow velocity of about 0.1 cm/s. Higher voltages result in more rapid separations but efficiency is degraded due to resistance to mass transfer within the mobile phase ("inter-micellar" and "intra-column") and eventually due to the formation of temperature gradients. Lower voltages result in long analysis times and a reduction in efficiency due to excessive axial diffusion.

In summary the MECC technique shows promise for the high efficiency separation of a variety of samples which are at least sparingly soluble in water. Fundamental studies of column selectivity are in progress in our laboatory. The results of these studies should provide insights for choosing suitable micelle/ buffer solutions for separating sample components which differ in specific molecular properties.

Acknowledgment

This research was sponsored jointly by the Division of Chemical Sciences, Office of Basic Energy Sciences, U. S. Department of Energy, under contract DE-ASO-83ER13127 with the University of Tennessee (Knoxville), the Office of Health and Environmental Research, U. S. Department of Energy, under contract DE-ACO-84OR21400 with the Martin Marietta Energy Systems, Inc., Oak Ridge, Tennessee, and under appointment to the Laboratory Participation Program administered by Oak Ridge Associated Universities for the U.S. Department of Energy.

Literature Cited

1. Terabe, S.; Otsuka, K.; Ichikawa, K.; Tsuchiya, A.; Ando, T. Anal. Chem. 1984, 56, 111-113.
2. Armstrong, D. W. Separation and Purification Methods 1985, 14(2), 213-304.
3. Jorgenson, J. W.; Lukacs, K. D. Anal. Chem. 1981, 53, 1298-1302.
4. Martin, M.; Guiochon, G.; Walbroehl, Y.; Jorgenson, J. W. Anal. Chem. 1985, 57, 559-661.
5. Terabe, S.; Otsuka, K.; Ando, T. Anal. Chem. 1985, 57, 834-841.
6. Otsuka, K.; Terabe, S.; Ando, T. J. Chromatogr. 1985, 332, 219-226.
7. Burton, D. E.; Sepaniak, M. J.; Maskarinec, M. P. J. Chromatogr. Sci., 1986, 24, 347-351.
8. Terabe, S.; Ozaki, H.; Otsuka, K.; Ando, T. J. Chromatogr. 1985, 332, 211-217.
9. Walbroehl, Y.; Jorgenson, J. W. Anal. Chem. 1986, 58, 479-481.
10. Burton, D. E.; Sepaniak, M. J.; Maskarinec, M. P. Chromatographia, in press.
11. Sepaniak, M. J.; Cole, R. Anal. Chem. submitted for publication.
12. Guiochon, G. Anal. Chem. 1981, 53, 1318-1325.

RECEIVED December 19, 1986

Chapter 7

Amphiphilic Ligands in Chemical Separations

E. Pramauro, C. Minero, and E. Pelizzetti

Dipartimento di Chimica Analitica, Università di Torino, Torino 10125, Italy

A series of 4-alkylamido-2-hydroxybenzoic acids containing a different number of carbon atoms in the alkyl-
amido group has been studied as model ligands for metal
ion extraction in aqueous micellar solutions of nonionic
surfactants. Their acid-base properties and reactivity
towards metal ions in the presence of micelles were investigated. By operating at a proper temperature, the
separation of the iron(III) chelate complexes into a
micellar rich phase was achieved and the extraction efficiency was correlated with the ligand hydrophobicity.

The use of organized molecular assemblies in analytical chemistry has
lead to the improvement of existing methods and to the development of
new procedures (1, 2). In particular, its applications in chemical
separations, including chromatography and extraction, seems to be
very promising (3, 4).

The phase separation of nonionic micellar solutions above the
cloud point has been succesfully applied to the liquid-liquid extraction of some metal chelate complexes (5, 6). In these systems the
concentration of the analyte takes place in the micellar rich layer,
which can be readily analyzed.

Although this approach can be interesting in analytical chemistry because it allows one to conduct extractions without using organic non miscible solvents, no systematic investigations were performed concerning the parameters which can regulate the efficiency of
the process, such as the effect of the ligand hydrophobicity, the variation of the chemical properties of reagents in the presence of micelles, the kinetics of complexation and extraction and so on.

In this work, some of the above mentioned features were investigated for a simple extraction model, using suitable complexing amphiphiles having different hydrophobicity.

0097-6156/87/0342-0152$06.00/0
© 1987 American Chemical Society

A series of compounds containing the same chelating moiety, namely 4-aminosalicylic acid, with different alkyl chains, was synthesized. The structure of the ligands (PAS-C$_n$) is the following:

$$R-CONH\!\!-\!\!\bigcirc\!\!-\!\!COOH \qquad C_n : C_{n-1}H_{2n-1}-CO$$

The micellar parameters were previously determined (7).

Since aggregation occurs for these compounds only at high pH values, a study of complexing properties of aggregates in the presence
of usual transition metal ions cannot be performed. At lower pH, however, the PAS-C$_n$ molecules can be readily solubilized in the presence of nonionic surfactants (e.g. Brij 35: polyoxyethylene(23)dodeca-
nol) and the obtained mixed micelles exhibit complexing capability
in acidic media.

In order to investigate the separation mechanism, the model system iron(III)-PAS-C$_n$ was chosen and its properties were studied in
the presence of nonionic micelles.

Experimental Section

Potentiometry. The dissociation constants of 4-alkylamido-2-hydroxy-
benzoic acids in the presence of Brij 35 micelles were measured at
25°C and 0.10 M ionic strength (NaNO$_3$). The ligand (0.002 M) was titrated with 1 M NaOH using a 655-Multi-Dosimat automated titrator
(Metrohm), equipped with a 605-pH-meter and a 614-Impulsomat unit.
The titrations were performed under N$_2$ flow, very slowly, in order to
allow the electrode equilibration.

Chromatography. The retention volumes of PAS-C$_n$ were measured with
a Perkin Elmer S-2 chromatograph, equipped with a UV-VIS-LC-55 B detector. A μ-Bondapak C$_{18}$ reverse phase column (Waters) was used.
Mobile phases containing Brij 35 (ionic strength: 0.10 M) were filtered through a 0.45 μm cellulose membrane filter (Millipore). Each
solute was dissolved in Brij 35 solutions before the runs; 5-10 μl
of the sample solution at a concentration in the range 0.001-0.003 M
were injected and the elution was performed at constant flow rate
(1-2 ml/min), at a fixed pH (2 or 6), at room temperature (25$\pm$ 1°C).
The absorbances were monitored at 280 nm.

Spectrophotometry. The absorbances of iron(III)-PAS-C$_n$ complexes in
the presence of Brij 35 micelles were measured at the wavelength of
the maximum (520 nm). Experiments were performed in the presence of
0.05 M HNO$_3$, at 0.10 M ionic strength (NaNO$_3$ was added), at 25°C.
The investigated surfactant concentration was in the range 0.001-0.01
M. The iron(III) present in the micellar rich phase in extraction

experiments was also measured spectrophotometrically, at 520 nm, after dilution of an aliquot of this layer with a buffered solution of
Triton X 100 (polyoxyethylene(9.5)-p-1,1,3,3-tetramethylbutylphenol)
2-5 % w/v, to ensure a cloud temperature enough high in order to
avoid turbidity effects during the measurements.

A Cary 219 spectrophotometer (Varian) was used throughout the
work.

Extraction. Extraction experiments were performed using suitable
nonionic surfactants or their mixtures having cloud point transition
temperatures not far from the room temperature. The surfactant concentration was in the range 1-5 % w/v.

The analyte content was 5-10 ppm, with a ligand concentration
in excess (ca. ten times with respect to the metal ion). The pH was
adjusted with a proper buffer (acetic acid / acetate or chloroacetic
acid / chloroacetate) and inert salt (NaNO$_3$) was added in order to
increase the density of the aqueous rich lower phase, which facilitates fast centrifugation.

Mixtures of Triton X 100 and BL 4.2 (polyoxyethylene(4.2)dodeca-
nol) were used in this part of the work.

The complex formation was fast in the reported conditions and,
after few minutes, the absorbance of the solution showed no changes.

After heating at a constant temperature (ca. 35°C), above the
cloud point of the mixture, the heterogeneous dispersion was centri-
fuged at 3400 r.p.m. for 15 min. A deep violet micellar rich upper
phase was then obtained. The centrifuge vessels were calibrated in
order to allow the measurement of the micellar phase volume; aliquots
of this layer were taking with a syringe for the analysis.

DC-Plasma Spectrometry. Some control measurements of the analyte
content in the aqueous extracted phase were performed using a Spec-
traspan IV apparatus (Spectrametrics). The emission line at 259.5 nm
was used.

Results and Discussion

Binding Constants of Ligands with Nonionic Micelles

The acid-base properties of the amphiphilic ligands change in the
presence of micellar aggregates due to the well known partition equilibrium of both the acid and anionic form. A continuous increase in
the apparent pK$_a$ was observed with increasing concentration of micellized surfactant (see Table I).

According to the simple pseudophase model of Berezin (8), the
binding constants between the ligands and the micelles have been calculated using the following equation:

$$K_{a(app)}^{-1} = K_{a(w)}^{-1} + K_{HA} K_{a(w)}^{-1} C_D \qquad (1)$$

where $K_{a(app)}$ is the dissociation constant of carboxylate group in the presence of Brij 35, $K_{a(w)}$ is the same constant in water, K_{HA} is the binding constant of the undissociated PAS-C$_n$ to the micelles and C_D is the concentration of micellized surfactant ($C_D = C_{tot} - CMC$). The critical micellar concentration for Brij 35, measured with the surface tension method is 6×10^{-5} M, in the experimental conditions.

Table I.   Values of $pK_{a(app)}$ for PAS-C$_2$, PAS-C$_4$ and PAS-C$_7$, at 25°C, I = 0.10 M, in the Presence of Brij 35 Micelles

Brij 35, $C_D \times 10^3$ (M)	Measured $pK_{a(app)}$		
	PAS-C$_2$	PAS-C$_4$	PAS-C$_7$
0.94	3.12	3.14	
1.44	3.14	3.20	
1.94	3.16	3.26	
2.44	3.20	3.32	
2.94	3.26	3.40	
3.94	3.29	3.47	
4.94	3.33	3.59	
6.94	3.55	3.68	
9.94	3.65	3.84	4.40
14.94			4.40
19.94			4.50

Plots of experimental data according to Equation 1, for PAS-C$_2$ and PAS-C$_4$, at the lower surfactant concentrations are shown in Figure 1.

Since the evaluation of this parameter is very important, it was also measured using the micellar HPLC technique (9), which allows a better estimation of the partition coefficients in the presence of quite high concentrations of surfactant. The chromatographic parameter P was measured for each ligand as a function of surfactant concentration, according to Equation 2:

$$P = P_{sw}^{-1} + K_{HA} P_{sw}^{-1} C_D \qquad (2)$$

where $P = V_s / (V_e - V_m)$, V_s and V_m are the volume of stationary and mobile phase, respectively, and V_e is the elution volume. P_{sw} represents the partition coefficient of solutes between the stationary and the aqueous phase.

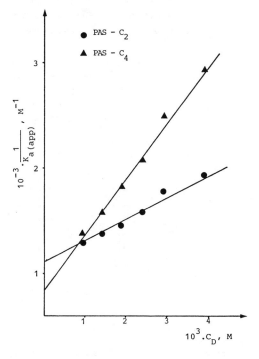

Figure 1. Plots of $1/K_{a(app)}$ as a function of micellized Brij 35 concentration, according to Equation 1, for PAS-C_2 and PAS-C_4.

For the investigated PAS-C$_n$ ligands, the binding constant for the undissociated form clearly increases with n, from 170 M^{-1} for C$_2$ and 500 M^{-1} for C$_4$ up to ca. 1500 M^{-1} for C$_7$, whereas for the anionic form it becomes significant only for C$_7$ (110 M^{-1}). The obtained data is in good agreement with the values estimated from pK$_{a(app)}$ shift, at low surfactant concentration, and allow us to define a minimum chain length for the ligand in order to have a strong binding to the micelles, both in acidic or anionic form.

Complex Formation Constant in the Presence of Micelles

The kinetics (10) and the complex formation equilibria in the presence of nonionic micelles have been also investigated, at constant acidity. The stoichiometry was assessed by using Job's method and the apparent stability constants were evaluated according to Frank and Ostwald procedure (11), as previously reported for the systems iron/sulfosalicylate and iron/salicylate in homogeneous aqueous acidic solution (12, 13).

For the equilibrium reaction:

$$Fe^{3+} + HSal^- \rightleftharpoons FeSal^+ + H^+$$

where HSal$^-$ indicates the dissociated chelating moiety of the ligand, the observed changes in the apparent formation constants (K$_c$) can be directly related with the variation of the apparent pK$_a$, previously discussed. For the less hydrophobic ligands, the increase of the surfactant concentration gave rise to higher K$_c$ values (e.g., by increasing the Brij 35 concentration from 0.001 M to 0.01 M, the observed change in K$_c$ was from 2.5x10^3 to 4.0x10^3 for PAS-C$_2$ and from 2.3x10^3 to 3.0x10^3 for PAS-C$_4$, respectively. Due to its lower solubility, PAS-C$_7$ was investigated in a narrow surfactant concentration range (0.01-0.02 M); a nearly constant K$_c$ value (ca. 3x10^3) was measured for this compound.

The experiments performed clearly showed that, whereas the complexation of iron(III) is not very much dependent on the ligand hydrophobicity, the association of the charged 1:1 chelates to the micelles and then the efficiency of the concentration process, markedly increases.

Extraction of Iron(III) from Micellar Solutions

The surfactant system Triton X 100 / BL 4.2 was chosen because its suitable cloud temperature range and the good solubilizing capability towards the ligands. Table II summarizes the properties of some investigated mixtures.

The analyte content, after extraction, was determined both in

the micellar and in the aqueous rich phase by VIS-spectrophotometry.
Calibration curves were made with micellar phases containing the dissolved ligands, in the absence of iron(III). To these solutions, separated by centrifugation, were added known amounts of analyte and
the absorbances were recorded.

The standard addition method was also applied to the extracted
micellar layers containing iron(III). The results obtained with both
procedures were found in good agreement, as well as which obtained
from DC-plasma spectrometry after analysis of the aqueous dilute phases.

The extraction efficiency was then calculated from at least four
independent measurements; the influence of the experimental parameters was also investigated.

Table II. Properties of Some Triton X 100/BL 4.2 Mixtures
in the Experimental Conditions (NaNO$_3$ 5 % w/v)

Triton X 100 : BL 4.2 ($\%$, w/v)	Cloud Temperature ($^\circ$C)	Volume of the Micellar Phase ($\%$)
0.50 : 0.50	26.3 - 26.5	6.5
0.75 : 0.75	26.6 - 26.9	8.5
1.00 : 1.00	26.8 - 27.0	11.0
1.25 : 1.25	26.8 - 27.1	14.3
1.50 : 1.50	27.0 - 27.3	21.1
2.00 : 2.00	27.5 - 27.7	30.6
2.50 : 2.50	27.8 - 28.0	31.0

Figure 2 shows the effect of the ligand hydrophobicity on the
analyte recovery, measured at pH 3.5, in the presence of Triton X 100
(1% w/v) and BL 4.2 (1% w/v); added NaNO$_3$: 5% w/v; iron(III): 1×10^{-4} M
and PAS-C$_n$: 2×10^{-3} M.

As it can be seen, quantitative recovery of iron(III) has been
obtained in the reported conditions using the more hydrofobic compounds (PAS-C$_{10}$ or higher analogues). However, the lower solubility
of these long chain molecules can limit the ligand concentration
available in the system, keeping the volume of micellar extraction
phase constant.

The extraction performances under different experimental conditions (i.e. varying the pH, the composition of surfactant mixtures,
the amount of chelating compound) were also investigated for our test
system. The results are shown in Table III.

All the extractions were performed in the presence of added salt
(NaNO$_3$, 5% w/v). The ligand concentration for the experiments performed at various pH was 2×10^{-3} M. For the runs in which surfactant

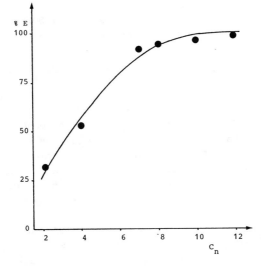

Figure 2. Plot of the percent recovery of iron(III) as a function of ligand alkyl chain length.

composition or ligand concentration were changed, the pH was constant
(3.5).

Table III. Extraction Efficiency as a Function of Experimental
Parameters for Iron(III)-PAS-C$_7$

pH	% E	PAS-C$_7$, M	% E	Triton X 100/ BL 4.2, % w/v	% E
2.00	39.0	$5.0x10^{-4}$	80.0	0.50:0.50	77.6
2.65	74.5	$1.0x10^{-3}$	87.5	0.75:0.75	93.6
3.10	88.4	$1.5x10^{-3}$	92.0	1.00:1.00	93.7
3.50	93.7	$2.0x10^{-3}$	93.7	1.50:1.50	94.0
3.75	94.3				

Conclusions

The results obtained with the reported extraction model showed that
the separation of charged species is possible, provided a suitable
ligand hydrophobicity. Further analytical developments of these mul-
tiphase extraction systems will require an accurate investigation of
the equilibria and kinetic processes occurring at the interfaces, as
well as the study of the micellar structure and properties of the
host aggregates.

Other functionalized surfactants having different complexing
groups and modular lipophilic chains, together with various nonionic
solubilizing surfactants, are presently under investigation in our
laboratories.

Acknowledgments

Support of this work by C.N.R (Rome) and European Standardization
Office, under Contract DAJA 45-85-C-0023, is gratefully appreciated.

Literature Cited

1. Hinze, W. L. In "Solution Chemistry of Surfactants"; Mittal,
    K. L., Ed.; Plenum Press: New York, 1979; p. 79.
2. Pelizzetti, E.; Pramauro, E. Anal. Chim. Acta 1985, 169, 1-29.
3. Cline Love, L. J.; Habarta, J. G.; Dorsey, J. G. Anal. Chem.
    1984, 56, 1133-48 A.
4. Armstrong, D. W. Separation and Purification Methods 1985, 14,
    213-304.
5. Watanabe, H.; Tanaka, H. Talanta 1978, 25, 585-9.

6. Watanabe, H. In "Solution Behavior of Surfactants: Theoretical and Applied Aspects"; Mittal, K. L.; Fendler, E. J., Eds.; Plenum Press: New York, 1982; Vol. II, p. 1305.

7. Pelizzetti, E.; Pramauro, E.; Barni, E.; Savarino, P.; Corti,M.; Degiorgio, V. Ber. Bunsenges. Phys. Chem. 1982, 86, 529-32.

8. Yatsimirskii, A. K.; Martinek, K.; Berezin, I. V. Tetrahedron 1971, 27, 2855-68.

9. Armstrong, D. W.; Nome, F. Anal. Chem. 1981, 53, 1662-6.

10. Pramauro, E.; Pelizzetti, E.; Cavasino, F. P.; Sbriziolo, C. in preparation.

11. Frank, H. S.; Oswalt, R. L. J. Am. Chem. Soc. 1947, 69, 1321-5.

12. Saini, G.; Mentasti, E. Inorg. Chim. Acta 1970, 4, 210-4.

13. Saini, G.; Mentasti, E. Inorg. Chim. Acta 1970, 4, 585-8.

RECEIVED October 24, 1986

Chapter 8

Coacervation of Polyelectrolyte-Protein Complexes

P. L. Dubin, T. D. Ross, I. Sharma, and B. E. Yegerlehner

Department of Chemistry, Indiana University-Purdue University at Indianapolis, Indianapolis, IN 46205

Complex formation between bovine serum albumin or ribonuclease and the cationic polymer poly(dimethyldiallylammonium chloride) was studied, with the eventual goal of using this phenomenon to separate globular proteins according to their surface charge density. Complex formation, generally accompanied by coacervation, occurs abruptly with increasing pH at constant ionic strength. This critical pH, signaled by a rapid increase in turbidity, increases with ionic strength. For BSA, the net protein charge at critical pH is a linear function of ionic strength. Turbidimetric titration curves generated by adding protein to polyion are very different from those obtained when the sequence of addition is reversed, an observation inconsistent with highly cooperative binding. Turbidimetric titrations conducted with mixtures of BSA and RNAse so far have not given clear evidence of selective coacervation, but optimization of ionic strength and polymer:protein stoichiometry may provide better separations.

Pairs of oppositely charged polyelectrolytes typically form complexes from aqueous solution (1). Depending primarily on the molecular weights and linear charge densities, these complexes may be amorphous solids (2), liquid coacervates (3-5), gels (7), fibers (7), or soluble (colloidal) aggregates (8-10). A special case of this phenomenon is the formation of complexes between globular proteins, away from their isoelectric points, and synthetic or natural polyelectrolytes of opposite charge. The formation of such complexes has been noted with regard to "colloid titration", in which protein concentrations can be determined by turbidimetric titration (11), or in the context of stabilization/entrapment of enzymatic proteins (12-14). Systems investigated to date include insulin/poly(vinylpyrrolidone-crotonic acid) (14), catalase/poly(2-(trimethylammonium)ethylmethacrylate) (13), penicillin amidase/poly(N-ethyl-4-pyridinium bromide) (12) and α-chymotrypsin/poly(N-ethyl-4-vinylpyridinium bromide) (12).

We have been studying complexes formed between polyelectrolytes and oppositely charged mixed micelles using turbidimetry (15-18), dynamic light scattering (18-20), viscometry (15,16), and dialysis/ultrafiltration (20). These concepts and methodologies, in our opinion, are closely relevant to an understanding of polyelectrolyte-protein interactions. We believe studies of this sort can shed insight on the coulombic interaction of proteins with DNA (21) or other biological polyelectrolytes, such as heparin. Furthermore, we believe that selective coacervation of proteins according to their isoelectric points could prove to be a novel method for protein separations, essentially unlimited in scale and relatively inexpensive. This last field is essentially unexplored, although its potential has not completely escaped notice (22).

Experimental

Poly(dimethyldiallylammonium chloride) (PDMDAAC) (1), a product of

1

Calgon Corp., trade name "Merquat-100", was used as is with no further purification. The nominal molecular weight (MW) is 2×10^5; absolute MW values, obtained by combined size exclusion chromatography and light scattering, are $\bar{M}_w = 5 \times 10^5$ and $\bar{M}_n = 2 \times 10^4$ (23). All proteins were obtained from Sigma Chemical Corp.

Turbidimetric titrations were conducted in three ways. In "Type 1" titrations, a mixture of PDMDAAC (0.04-4.0 g/L) and protein (0.1-10 g/L) were combined at pH $\leqslant 4$ in distilled deionized water or dilute (0.05-0.5M) NaCl. The optical probe (2 cm path length) of a Brinkman PC600 probe colorimeter (240 nm), and a combination pH electrode connected to an expanded scale pH meter (Orion 811 or Radiometer pH M26), were both placed in the solution. Titrant (0.50 M NaOH or 0.50 M HCl) was delivered from a 2.0 mL microburet (Gilmont) with gentle stirring. Alternatively, turbidity was monitored while a protein solution was added to PDMDAAC (at constant ionic strength) or vice-versa. Turbidity was reported as 100-%T, which is linearly proportional to the absolute turbidity in the range 80<%T<100.

Results and Discussion

Typical results for "Type 1" turbidimetric titrations are shown in Figure 1, for BSA and RNAse, in the presence of 0.04 wt. % PDMDAAC at ionic strength 0.2 M.

A number of features of these data merit attention. First, the rise in turbidity is rather abrupt, taking place within ca. 0.3 pH units. Thus, there appears to be a well-defined critical pH below which no association is observed. Turbidity values in the range 100-%T<30 are stable for at least several hours and curves such as those in Figure 1 are readily reversible, both observations suggestive of an equilibrium system. Turbidity appears to result from coacervation: centrifugation at

3000 rpm for 30 min produces two liquid phases, both optically clear, the denser phase obviously rich in macromolecular solute. The more basic protein, RNAse, exhibits a higher pH_{crit}, expected on the basis of its lower net negative charge at any pH.

Data for "Type 1" titrations of BSA over a wide range of ionic strengths I are assembled in Figure 2, as pH_{crit} vs. I. Despite considerable scatter, one may observe a decrease in slope in the range $7 < pH < 9$. It is of interest to note that the dependence of the net charge, Z, of BSA on pH is largest outside of this pH region as shown in Figure 3. When the critical value of Z is plotted as a function of ionic strength, as in Figure 4, the resulting phase boundary is virtually linear.

Our interpretation of Figure 4 is as follows. Complex formation requires that some sequence of polyion residues cooperatively ion-pairs with carboxylate groups on the protein surface. At low pH, the surface density of such groups is small and any polyion-protein ion-pairs are transient. At high pH, protein surface carboxylate density is sufficiently large to facilitate multiple attachment of the polymer chain to the protein surface. The addition of simple electrolyte opposes such interactions by electrostatic screening. Consequently, a high negative surface charge density (corresponding to a higher pH) is required for complex formation at higher ionic strengths. While the foregoing discussion focusses on the interaction of a single protein with a small sequence of polyion residues, one presumes that many proteins can bind to a single polymer chain (effective hydrodynamic diameter ca. 400 Å) above critical conditions. If the polyion-protein complex achieves electroneutrality, with a loss of counterions, further higher order association may lead to bulk phase separation.

A second type of turbidimetric titration yields information on the stoichiometry of complex formation. Figure 5 shows the results of titrating 0.01 wt. % PDMDAAC with 0.01 wt. % BSA in pure water (initial pH 6.8), i.e. conditions at which complex formation should take place. The absence of significant turbidity in the first half of the titration could indicate that protein molecules are dispersed among polyelectrolyte chains, i.e. in a non-cooperative manner, and thus no highly scattering aggregates are formed. The sharp maximum suggests a stoichiometric point, at which complexes achieve charge neutralization and consequently aggregate further to form coacervate. Since the initial volume of polymer solution was 5.0 mL, an "end-point" at 150 mL of added BSA suggests that the stoichiometric ratio for BSA:PDMDAAC is 30:1 (wt. basis) or 90:1 (mole basis). In view of the contour length of the polyion, on the order of 5000 Å, and its formal net charge of +1200, it is not unreasonable to contemplate the binding of ca. 90 molecules of BSA per polymer chain. Binding of more protein, beyond this stoichiometric point, might result in charge reversal of the complex, thus destabilizing the coacervate phase and reducing the turbidity. Titrations of this type at varying initial polymer concentrations need to be done to reveal whether the turbidity maximum occurs at a well-define protein:polymer stoichiometry.

The titration of protein with polymer, shown in Figure 6, exhibits very different features than the titration curve of Figure 5. Turbidity appears at once and, initially, exhibits a linear dependence on PDMDAAC concentration. The rate of turbidity development subsequently increases about four-fold, and then diminishes slightly. We may speculate that, in the presence of excess protein, all polyions

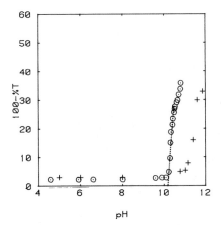

Figure 1. "Type 1" turbidimetric titrations for BSA (O) and RNAse (+), both 1 g/L, in the presence of 0.4 g/L PDMDAAC and 0.2 M NaCl. Broken line shows evaluation of critical pH.

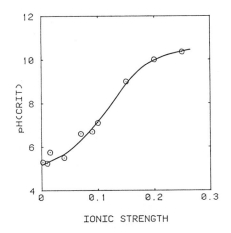

Figure 2. Critical pH vs. ionic strength for 1 g/L BSA + 0.4 g/L PDMDAAC, from titrations such as shown in Figure 1, conducted in Na Cl solutions of varying molarity.

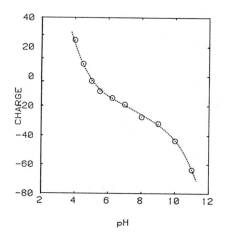

Figure 3. pH titration curve for BSA in 0.15 M NaCl relative to point of zero charge. Reprinted with permission from ref. 24. Copyright 1955 American Chemical Society.

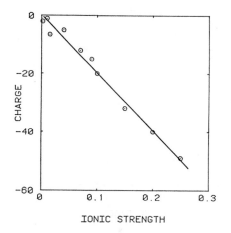

Figure 4. Protein net charge at critical conditions for complex formation as a function of ionic strength; from data in Figure 2.

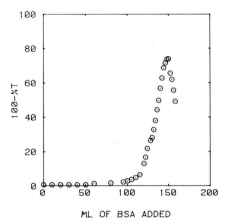

Figure 5. Titration of 0.1g/L PDMDAAC with 0.1 g/L BSA in pure water at pH 7±0.5. Initial volume 5.0 mL.

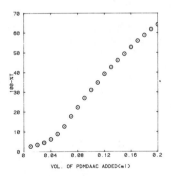

Figure 6. Titrations of 1 g/L BSA with 1 g/L PDMDAAC in pure water at pH 7±0.5.

added initially produce equivalent complexes so that the scattering is proportional to polymer concentration. Subsequently, the formation of coacervate can lead to a marked increase in turbidity which is not simply related to the number of primary (i.e. intrapolymer) complexes. We note that these titration curves do not produce a well-defined end-point, as previously reported for this same system (11).

As shown in Figure 1, the critical pH values for BSA and RNAse in 0.2 M NaCl are separated by about one pH unit. When a solution containing BSA and RNAse, each at 0.5 g/L, is titrated in like fashion, a single turbidity curve is observed at an intermediate pH. At this point, it is not clear whether both proteins are participating jointly and simultaneously in complex formation, and therefore not being separated. If the titration is stopped, say at 100-%T = 20, the coacervate may be removed and the titration of the supernatant continued. Analysis of these successive coacervate fractions is currently underway, employing size exclusion chromatography to determine the concentrations of the two proteins.

Conclusions and Future Work

Complex coacervation of globular proteins with the strong polycation PDMDAAC appears to be a purely coulombic phenomenon, controlled by ionic strength and protein net charge (via pH). Turbidimetric pH titrations suggest that the process resembles a critical phenomenon since solutions go from optical clarity to high turbidity over a pH interval of 0.3 units or less.

Critical conditions for BSA and RNAse in 0.2 M NaCl are separated by about one pH unit. However, co-titration of the two proteins does not produce titration regions identifiable with the individual proteins. Analyses are in progress to determine whether the two proteins are indeed co-precipitating. If this is the case, separation may be improved in one of several ways. First, the critical pH's are better separated at lower ionic strengths under which conditions mixed titrations will be carried out. Secondly, the turbidimetric titrations, which exhibit measurable breath, may be made more narrow by using a narrow molecular weight (MW) distribution PDMDAAC; the very large poly-dispersity of the current material could give rise to a range of pH_{crit} values for a single protein even with only modest MW dependence of pH_{crit}. Lastly, we can point out that the stoichiometry noted in our discussion of Figure 5 above suggests that the "Type 1" titrations presented here involve a large excess of PDMDAAC. Greater protein selectivity might be found closer to stoichiometry.

Acknowledgments

Support from the National Science Foundation (Grant #DMR 8507479), the Petroleum Research Fund (Grant #17656-B7-C), and a University "Project Development Program Small Grant" is gratefully acknowledged.

References

1.   Tsuchida, E.; Abe, K. "Interactions between Macromolecules in Solution and Intermacromolecular Complexes," Advances in Polymer Science, vol. 45, Springer-Verlag, Berlin, 1982.

2. Michaels, A.S.; Miekka, R.G. J. Phys. Chem. 1961, 65, 1765.

3. Bungenberg de Jong, H.G., in "Colloid Science"; H.R. Kruyt, Ed.; Elsevier, Amsterdam, 1949, vol. II., Chapter X.

4. Veis, A.; Bodor, E.; Mussell, S. Biopolymers 1967, 5, 37.

5. Polderman, A. Biopolymers 1975, 14, 2181.

6. Tsuchida, E.; Osada, Y.; Abe, K. J. Polymer Sci. 1976, 14, 767.

7. Tsuchida, E.; Abe, K; Honma, M. Macromolecules 1976, 9, 112.

8. Kabanov, V.A.; Zezin, A.B. Macromol. Chem. Suppl. 1984, 6, 259.

9. Dauzenberg, H.; Linow, K.-J.; Philipp, B. Acta Polymerica 1982, 33, 619.

10. Shinoda, K.; Sakai, K.; Hayashi, T.; Nakajima, A. Polymer J. (Japan) 1976, 8, 208.

11. Kokufuta, E.; Shimiza, H.; Nakamura, I. Macromolecules 1982, 15, 1618.

12. Margolin, A.L.; Sherstyuk, S.F.; Izumrudov, V.A.; Zezin, A.B.; Kabanov, V.A. Eur. Polymer J. 1985, 146, 625.

13. Kohjiya, S.; Maeda, K.; Ikushima, Y.; Ishihara, Y.; Yamashita, S. Nippon Kagaka Kaishi 1985, 12, 2302.

14. Glikina, M.B.; Kuznetzova, N.P.; Nemtzova, N.N.; Zhukovakaya, L.L.; Drozdova, E.V.; Samsonov, E.V. 15th Conference on Synthesis, Structure and Properties of Polymers, Soviet Academy of Science, April, 1968, p. 221.

15. Dubin, P.L.; Oteri, R. Polymer Preprints 1982(1), 23, 45.

16. Dubin, P.L.; Oteri, R. J. Colloid Interface Sci. 1983, 90, 453.

17. Dubin, P.L.; Davis, D.D. Colloids Surfaces 1985, 13, 113.

18. Dubin, P.L.; Rigsbee, D.R.; McQuigg, D.W. J. Colloid Interface Sci. 1985, 105, 509.

19. Dubin, P.L.; Davis, D.D. Macromolecules 1984, 17, 1294.

20. Rigsbee, D.R.; Dubin, P.L.; Fallon, M.A. accepted for publication.

21. Lohman, T.M.; De Haseth, P.L.; Record, M.T., Jr. Biophys. Chem. 1978, 8, 281.

22. Michaels, A.S. Chem. Eng. Prog. June, 1984, p.19.

23. Getman, G., Calgon Corp., private communication.

24. Tanford, C.; Swanson, S.A.; Shore, W.S. J. Am. Chem. Soc. 1955, 77, 6414.

RECEIVED April 10, 1987

Chapter 9

Extraction of Proteins and Amino Acids Using Reversed Micelles

T. Alan Hatton

Department of Chemical Engineering, Massachusetts Institute of Technology, Cambridge, MA 02139

The solubilisation of proteins and amino acids in organic
solvents by reversed micelles provides a new method for the
selective recovery, separation and concentration of
bioproducts using liquid¬liquid extraction techniques.
Selectivity is affected by electrostatic interactions between
the charged residues or moieties of the solute and the
surfactant headgroups. These interactions are mediated by
electrostatic screening as affected by solution ionic
strength. The more hydrophobic the amino acid residue, the
more favourable is the solubilisation of this residue in the
partially structured water pool of the reversed micelle
relative to the bulk, unstructured water phase.

The large scale recovery of polar biological molecules such as
enzymes and other proteins, nucleic acids, polypeptides and amino
acids has been given new emphasis in recent years as a result of the
significant advances made in recombinant DNA techniques and genetic
engineering. These advances have opened up many new vistas for the
practical utilisation of biotechnology on an unprecedented scale.
The problem still remains, however, as how best to achieve the
required separation, concentration and purification of the
biological products on a continuous basis. Many techniques are
available for this purpose, and each has its niche in the complex
arena of bioproduct separations. In some cases there may be more
than one separation method that would work, although generally these
methods were developed for bench¬scale separations and suffer from
loss of resolution on scale¬up. Moreover they do not offer the
advantages of economy of scale so frequently found with separation
operations in other chemically¬related industries.
A separation operation that has not received the consideration
it deserves for bioproduct recoveries is that of liquid¬liquid
extraction, which can offer both moderate to high selectivity, and
can be operated on a continuous, large-scale basis. One of the
primary reasons for this neglect is the lack of suitable solvent
systems having the desired selectivity and capacity for the products
of interest. Many of the products are ionic in character at the pH

conditions typical of fermentation media and tend to be insoluble and/or labile in the organic solvents traditionally used in extraction operations. As a consequence, organic solvents have not been considered likely candidates for these separations, and the field has not received the attention given to other techniques such as chromatography and ion-exchange.

In this paper, we consider a new class of extractants which do not have the inherent limitations of conventional solvents in the recovery of biological products from their native soups. The concepts are based on the observation that certain surfactants aggregate in apolar media to form reversed micelles (1). These micelles generally consist of a polar core which includes the surfactant headgroups, solubilised water, and any other solubilised polar materials, surrounded by hydrophobic surfactant tails protruding into the continuous organic medium enveloping the micelle. Under the appropriate conditions, these reversed micelle-containing organic solutions can exist in equilibrium with a bulk aqueous phase, and can be used to extract polar materials from this aqueous solution. The materials solubilised in the inner cores of these reversed micelles are generally water and salts, although other polar species, in particular proteins and amino acids, can also be taken up by these micelles and thereby solubilised in an otherwise inhospitable organic environment. The solubilisation of a protein in a reversed micelle is illustrated schematically in Figure 1.

In the development of these concepts for bioproduct recovery, it is important to have a good understanding of the factors influencing the selectivity of the extraction process for the solutes of interest. This is the topic of the overview presented here, where it will be apparent that electrostatic interactions are an important factor in this selectivity, but also that hydrophobic interactions can play a significant role in determining solubilisation behaviour.

Protein Solubilisation

The ability of proteins to transfer from an aqueous solution to a reversed micelle-containing organic phase, and be subsequently recovered in a second aqueous phase, was first established by the group of Luisi (2,3). It has since been suggested by van't Riet and Dekker (4,5) and Goklen and Hatton (6-9) that this phenomenon be exploited in the recovery, separation and concentration of bioproducts from complex aqueous mixtures. In the past three years, significant progress has been made in this direction, and it has been established that these solvents can be selective in the separation of binary and ternary protein mixtures (7-9) and in the recovery of an extracellular alkaline protease from a clarified fermentation broth (10). It has also been demonstrated that the process can be operated on a continuous basis (5).

The physico-chemical interactions that can be exploited in the selection and optimisation of these processes are discussed below. We will reserve for future communications a detailed discussion of the technological aspects and potential problems in the implementation of reversed micellar extraction of bioproducts in large-scale continuous operations, these topics being beyond the scope of this overview.

Effect of pH. The pH of the solution will affect the solubilisation
characteristics of a protein primarily in the way in which it
modifies the charge distribution over the protein surface. With
increasing pH, the protein becomes less positively charged until it
reaches its isoelectric point, or point of zero net charge, pI. At
pH's above the pI the protein will take on a net negative charge.
If electrostatic interactions play a significant role in the
solubilisation process, solubilisation with anionic surfactants
should be possible only at pH's less than the pI of the protein,
where the protein is positively charged and electrostatic
attractions between the protein and the surfactant headgroups are
favourable. At pH's above the pI, electrostatic repulsions would
inhibit the protein solubilisation. The reverse trends would be
anticipated in the case of cationic surfactants.

The expected trends are born out for the low molecular weight
enzymes ribonuclease-a, cytochrome-c, and lysozyme, as shown in
Figure 2. These results are presented as the percentage of the
protein transferred from a 1 mg/ml aqueous protein solution to an
equal volume of isooctane containing 50 mM of the anionic surfactant
Aerosol OT, or AOT (di-2-ethylhexyl sodium sulfosuccinate). As
anticipated, only at pH's lower than the pI was there any appreci⁻
able solubilisation of a given protein, while above the pI the
solubilisation appears to have been totally suppressed. Note,
however, that as the pH was lowered even further, there was a drop
in the degree of solubilisation of the proteins. This was
accompanied by the formation of a precipitate at the interface
between the two phases, attributed to a denaturation of the protein.

These results were not always found for larger proteins at the
same level of surfactant loadings. For the class of proteins
trypsin, alpha-chymotrypsin, elastase and alpha-chymotrypsinogen it
was found that the pH had to be reduced to values significantly
below the pI for there to be any appreciable solubilisation. Even
then the solubilisation occurred only over a very narrow pH range
before decreasing rapidly again with further decreases in the pH of
the aqueous feed phase, accompanied by precipitation at the
interface. With increased surfactant concentrations, the earlier
behaviour was recovered, i.e., the pI again marked the point of
transition between significant solubilisation and no solubilisation
of the protein (Figure 3).

These results all point to the electrostatic interactions
between the solute particles and the surfactant headgroups being a
controlling factor in the solubilisation process. While this is
undoubtedly true, results presented below on the effects of ionic
strength indicate some more subtle phenomena come into play, too.
The question that arises is precisely how is the protein positioned
within the micelle. The presence of reversed micelles in these
studies is undisputed. Measurements of water contents using
Karl-Fisher titration, and micelle size determinations via dynamic
light-scattering (11) and small-angle neutron scattering (12)
confirm their existence. This of course does not guarantee that the
protein is contained within the polar core of the micelle. Rather,
it is possible that solubilisation occurs by simple ion-pairing

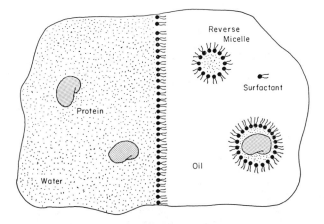

Fig. 1   Schematic Representation of Protein Solubilisation in Reversed Micelles.

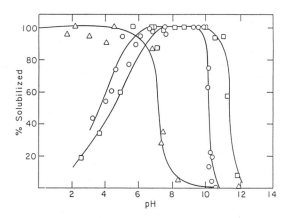

Fig. 2   Effect of pH on Solubilisation of Cytochrome-c (O), Lysozyme (□), and Ribonuclease-a (Δ).

between the positively charged surface residues and the anionic
surfactant headgroups. This scenario, however, ignores the fact
that over the pH range of interest, many of the surface residues
will be negatively charged, and would need to be hydrated when the
protein is pulled into the organic phase. This would point to
solubilisation within the water pool of the micelle.

The results discussed above relate to the anionic surfactant
AOT. For cationic surfactants, a different trend is observed.
Van't Riet's group have found that with trioctylmethyl ammonium
chloride as the surfactant, significant solubilisation of the enzyme
alpha-amylase was observed over a narrow pH range only, in the
vicinity of 10 ≈10.5 (4). In this pH range, it can be anticipated
that all basic residues will be deprotonated, the only charged
residues being the carboxyl groups bearing a negative charge. These
would be available for ion-pairing, and indeed it would appear that
this is the mechanism for the solubilisation of this protein.
Similar results have been observed in our laboratory, even for
proteins having low pI's, such as carbonic anhydrase and
ribonuclease-a. Hinze has also observed this trend (13).

Thus, it can be argued that depending on the type of surfactant
used, different solubilisation mechanisms could be operative. For
the anionic surfactant AOT, it appears that micellar solubilisation
is occurring, while ion-pairing is the mechanism for cationic
surfactants. It should be emphasised, however, that these general-
isations are not always valid. For instance, recent results on the
solubilisation of catalase using the cationic surfactant DTAB
(dodecyl trimethyl ammonium bromide) in n-octane, with hexanol as
cosurfactant, indicated that the solubilisation was by reversed
micelles and not by strict ion-pairing. While below the pI of about
5.3 there was no solubilisation of this enzyme, above this pH value,
significant transfer occurred. This result, plus the observation
that the micelle size increased dramatically above the pH of 5.3,
give strong evidence for the micelle solubilisation mechanism being
operative.

From these observations, a clearer picture of the solubilisation
process is beginning to emerge, although it is still not
sufficiently advanced to be applied in a predictive sense.

Effect of Ionic Strength. The effect of ionic strength is primarily
to mediate the electrostatic interactions between the protein
surface and the surfactant headgroups. The well-known Debye
screening determines the electrical double layer properties adjacent
to any charged surface, and affects the range over which
electrostatic interactions can overcome thermal motion of the solute
molecules. The characteristic distance for these electrostatic
interactions is the Debye length, which is inversely proportional to
the square root of the ionic strength (14). Thus, increases in the
ionic strength will decrease this interaction distance, and hence
inhibit the solubilisation of the protein. This decreased
interaction has been neatly confirmed in the case of AOT reversed
micelles in isooctane in equilibrium with salt solutions. As the
salt concentration increases, the repulsive headgroup interactions
between surfactants will be suppressed, permitting the formation of

smaller micelles, a trend which is evident in the results shown in Figure 4. The strong linear dependence of micelle size on the reciprocal square root of the ionic strength argues in favour of the importance of Debye screening in these systems.

The results shown in Figure 5 further confirm the importance of ionic strength effects on protein solubilsation in reversed micelles. With increasing ionic strength, there is a fairly abrupt change in the solubilisation, occuring at different salt concentrations for the different proteins. It is of interest to note that the order of appearance of these curves in this figure is different from that observed for the same proteins in the pH variation case. This degree of discrimination between the similarly-sized proteins was unexpected, and points to the sensitivity of the micelle extraction process to the individual structural features of each protein. At this stage no definite conclusions can be drawn as to the precise phenomena operative in this process, although it can be speculated that it relates to the surface topology of the protein, and in particular to the distribution of charged residues and hydrophobic patches over the surfaces of the protein.

The salt type will also affect the solubilisation of the proteins (11,15). Goklen and Hatton (9) observed that the importance of the salt effects in determining water solubilisation capacities in reversed micelles followed the now-classic lyotropic, or Hofmeister series within any given valency group. This points to the importance of the ion solvation effects, and possibly also specific adsorption phenomena in the Stern layers of the micelle wall and the protein surface (14) in these systems, again illustrating the dominance of electrostatic interactions in the solubilisation process. In addition, it was found that with $CaCl_2$ the range of pH's over which significant solubilisation occured extended beyond the pI of the protein, indicating some specifc ion binding of the divalent Ca^{++} cation to the protein thus modifying its charge characteristics and shifting the effective pI of the protein to higher values.

The effect of ionic strength is further evident in the results shown in Figure 6, where the degree of cytochrome-c solubilisation as a function of pH is given for two different KCl concentrations. Increasing the salt concentration results in a narrowing of the solubilisation peak to the extent that solubilisation occurs only at pH's significantly below the nominal pI of the protein. This could be attributed to either a simple Debye screening of the electrostatic interactions between the protein and the surfactant headgroups, which is overcome by the increased net positive surface charge of the protein at the lower pH's, or to specific or non-specific chloride ion binding to the protein, rendering it more negatively charged than would be indicated by its pI.

Whatever the reason for the effect of ionic strength on the solubilistion characteristics for these proteins, it is apparent that manipulation of ionic strength can be employed to vary the pH range over which solubilisation occurs, allowing greater selectivity in the extraction process.

Affinity Partitioning. A new area that has only just begun to be

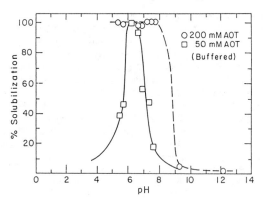

Fig. 3   Effect of pH and Surfactant Concentration on Solubilisation of Alpha-Chymotrypsinogen.

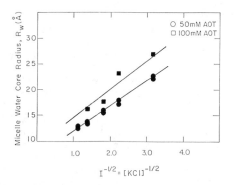

Fig. 4   Effect of Ionic Strength on Equilibrium Micelle Size for Protein-Free Solutions as Indicated by SANS

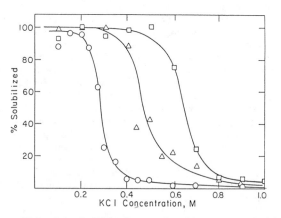

Fig. 5   Effect of KCl Concentration on Solubilisation of Cytochrome-c (O), Lysozyme (□), and Ribonuclease (Δ).

explored is the possibility of enhancing selectivity in reversed
micellar bioseparations through suitable synthesis of surfactants
endowed with molecular recognition capabilities (16). For instance,
the affinity ligands conventionally used in affinity chromatography
can be modified through the attachment of a long alkyl tail to the
active recognition group (frequently a substrate or inhibitor for
the enzyme). The tail length should be selected to ensure the
affinity surfactant straddles the reversed micelle surfactant shell,
permitting the head group to protrude into the polar core where it
can interact selectively with the proteins to be recovered, while at
the same time ensuring that the affinity group is anchored within
the extractant phase and is not stripped into the aqueous phase.
The basic concepts are illustrated schematically in Figure 7. Note
that for this concept to work, there is a need for only one affinity
surfactant per micelle, on average, since it is reasonable to assume
single occupancy for these protein–micelle complexes, and thus low
concentrations of these surfactants will be required relative to
those used for forming the reversed micelles.

Solubilisation of Amino Acids

An important class of biologicals is the amino acid group, not only
because they are valuable in their own right as feed supplements,
etc., but also because they are the substrates for the synthesis of
di- and oligo-peptides used, for example, as therapeutic and
analgesic drugs and artificial sweeteners, and are the building
blocks for all proteins. In fact, it is this latter point that
makes the study of amino acid solubilisation in reversed micelles
particularly interesting. A quantification of the interactions,
both electrostatic and hydrophobic, between the well-characterised
amino acids and the reversed micelle polar cores should provide
additional insight into the importance of these phenomena in protein
solubilisation experiments. Because of their molecular structure
and ionisable groups, at physiological pH conditions they are not
readily extracted into organic solvents using conventional ion-
pairing or chemical complexation techniques as found in e.g. the
hydrometallurgical industries.
    The general molecular structure of amino acids is illustrated
schematically in Figure 8. It is the presence of both the amino
group ($-NH_2$) and the carboxyl group ($-COOH$) attached to the
alpha-carbon that imparts to the amino acids many of their
interesting characteristics. In addition to these groups, the
residue, or "R-group," attached to the alpha-carbon is what
distinguishes one amino acid from another, and is responsible for
the unique physical properties of each species. These moieties can
be either cationic or anionic, or can be neutrally polar or
nonpolar, depending on the species considered (17). It is of
interest to note that under physiological conditions, both the amino
and the carboxyl groups will be ionised, so that, for nonionic
residues, the amino acids will be electrically neutral. It is this
zwitterionic character that makes for difficult separations by
extraction methods. For the charged residues the problem becomes
more acute.
    Solubilization of amino acids in AOT/isooctane reversed micelle

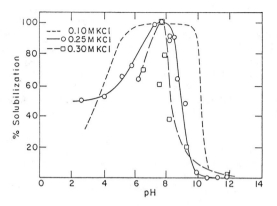

Fig. 6   Effect of pH and Ionic Strength on Solubilisation of
            Cytochrome-c.

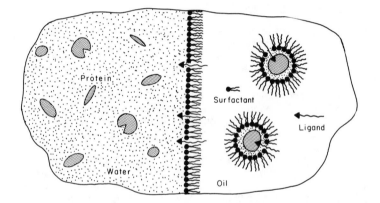

Fig. 7   Principle of Affinity Partitioning in Reversed
            Micelles.

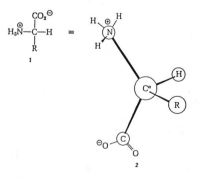

Fig. 8   Structure of Amino Acids.

solutions has been investigated in detail by Thien et al. (18). Extraction curves are shown in Figure 9 for a cationic and a neutral aromatic amino acid. Superimposed on the graphs are the titration curves calculated for the individual species. It is readily apparent from these data that the solubilisation is possible only when the amino acids bear a net positive charge, and that the degree of solubilisation is governed by the value of this net positive charge. For negatively charged species, solubilisation is suppressed. These results suggest that the solubilisation is governed primarily by electrostatic interactions between the anionic surfactant headgroups and the positively charged amino acid moieties. It is of interest to note that at low pH conditions the protonated amino group participates in the electrostatic interactions, but that at higher pH's, where the carboxyl group is deprotonated and bears a negative charge, this interaction is neutralised by the zwitterionic nature of the alpha-carbon group.

Thien et al. (18) also showed that the degree of solubilisation depends on the relative hydrophobicities of the amino acid residues. A measure of this effect is the slope of the solubilisation curve when replotted as a function of the net solute charge, as shown in Figure 10 for arginine. It was found that this slope correlated well with the hydrophobicity scale proposed by Bull and Breese (19), as is evident from Figure 11. It is intriguing to note that the more hydrophobic the residue the greater the solubilisation of the amino acid. This could be due to one of three effects, as discussed below.

The first possibility is that the solubilisation occurs by simple ion-pairing between the protonated amino groups and the surfactants. While this cannot be ruled out at low pH, it is not the operative mechanism at the intermediate pH's, where the deprotonated carboxyl groups for the charged polar amino acids must be hydrated and consequently the solute must be solubilised within the polar core of the reversed micelle itself. In addition, dynamic light scattering studies and Karl-Fischer water titrations do not indicate the significant decrease in water solubilisation that would be expected if the surfactant were tied up in ion pair formation and were no longer available for micelle formation.

A second scenario is that the nonpolar residue of the amino acid is located within the surfactant shell, permitting the strong interaction between the protonated amino group and the surfactant head required for the solubilisation of the solute. Again, this could not be the case for the charged polar amino acids. In addition, such an effect is not consistent with the Karl-Fischer titrations and light scattering experiments, which indicate an unchanged apparent surfactant head coverage on incorporation of the solute within the micelles. An increase in this value would be expected if the solute were to occupy part of the interfacial shell of the reversed micelle. Moreover, there is other evidence that in normal micelles, aromatic solutes such as benzene and naphthalene are not taken up in the apolar hydrocarbon core, but are rather located in the aqueous phase near the surfactant headgroups (20,21). This supports the contention that the neutral aromatic amino acids such as phenylalanine and tryptophan are solubilised totally within the water pools of the reversed micelles.

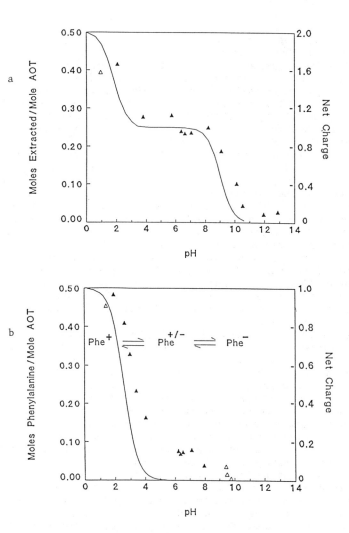

Fig. 9   Solubilisation of (a) Arginine, and (b) Phenylalanine
           in Reversed Micelles. The Solid Lines are Calculated
           Titration Curves Based on the pK_a's for the Amino
           Acids.

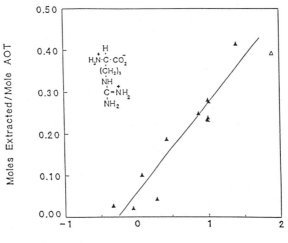

Fig.10   Solubilisation of Arginine as a Function of Net Charge.

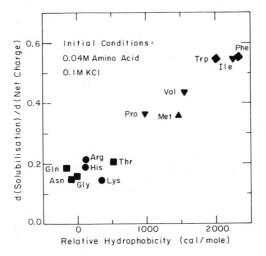

Fig.11   Correlation of the Slope of the Solubilisation vs. Net Charge Curve with the Relative Hydrophobicity of the Amino Acid Residue.

These observations beg the question: why do the hydrophobic solutes favour the polar cores of the reversed micelles more than do the polar amino acids? This effect would seem to be due to the unusual properties of the water within the reversed micelles, which is known to be more structured than water in the bulk aqueous phase (22). There is thus a distinct entropic advantage in removing the hydrophobic residue from the bulk aqueous phase, where it induces entropically disfavoured structural changes in this aqueous phase, and positioning it within the already partially-structured water within the micelle pool. The gain in entropy in such cases would be greater for the nonpolar residues than for the polar moieties, with the result that the former would be solubilised to a greater extent than the latter, as has been observed experimentally. This is essentially a manifestation of the well-known hydrophobic effect (23).

These results on amino acid solubilisation in reversed micelle solutions have indicated clearly that such systems could be useful for the recovery, separation and concentration of small, charged biological molecules from aqueous media. Furthermore, they have shed some light on the role that hydrophobic interactions will play in the solubilisation of more complex molecules such as proteins, which have a distribution of polar and nonpolar amino acid residues over their surfaces.

Conclusion

The use of reversed micelles in the selective recovery and concentration of low and high molecular weight bioproducts from dilute aqueous streams appears to be a promising new avenue for innovative research and applications. To date, it has been shown that electrostatic interactions between the charged solute residues and the surfactant headgroups, as well as hydrophobic effects, can play a significant role in determining the selectivity of this process for one protein over another. Moreover, there appears to be some latitude in the selection of surfactants and cosurfactants that enables enhancements in selectivity to be made over and above those already inherent in the process.

While some understanding of the factors responsible for the selectivity of the extraction process has been gained, the field is still in its infancy, and the challenge is to delve further into the subtleties of micellar solubilisation to obtain a more complete molecular level description of the electrostatic and hydrophobic interactions leading to the cooperative formation of the solute-solubilisate complexes. In this context, an area that has received no attention to date is the mechanistic description of the interfacial transport of the solutes, and only fragmentary data are available on mass transfer rates in these systems. Such information will certainly be required as the technique progresses from the exploratory stage to final applications in large scale production operations.

In conclusion, it can be stated that reversed micelles present an attractive alternative to conventional bioseparation methods, but that much fundamental and applied research is required before this potential can be fully realized. The intellectual, and possibly economic, rewards to be gained in the pursuit of these goals are adequate justification for a concerted effort in this area.

Acknowledgments

I am indebted to past and present graduate students Kent Goklen and Mike Thien, respectively, and undergraduates Karen Lee and Kim Thompson, for producing most of the results discussed in this overview. I have also drawn on some unpublished results by graduate students Andy Bommarius and Reza Rahaman, and I thank them for their contributions. This work was supported by the NSF Biotechnology Process Engineering Centre at MIT, the W.R.Grace Company, Alfa Laval, and a Nestle/Westreco Fellowship Grant.

Literature Cited

1. Fendler, J. H. "Membrane Mimetic Chemistry"; Wiley: New York, 1982.
2. Luisi, P. L.; Bonner, F. J.; Pelligrini, A.; Wiget, P.; Wolf R. Helv. Chim. Acta 1979, 62, 740-53.
3. Wolf, R. Ph.D. Thesis ETH#7027, Swiss Fed. Inst. Tech., Zurich, 1982.
4. Van't Riet, K.; Dekker, M. Proc. 3rd Eur.Cong. Biotechn., Vol. III, 1984, p.541.
5. Dekker, M.; Van't Riet, K.; Weijers, S. R.; Baltussen, J. W. A.; Bijsterbosch, B.; Laane, C. Biochem. Eng. J. 1987, in press.
6. Goklen, K. E.; Hatton, T. A. Biotech. Prog. 1985, 1, 69-74.
7. Goklen, K. E.; Hatton, T. A. Proc. ISEC '86, Vol. III, 1986, p.587.
8. Goklen, K. E.; Hatton, T. A. Sep. Sci. Technol. 1987, in press.
9. Goklen, K. E.; Hatton, T. A. Biotech. Bioeng. 1987, submitted.
10. Rahaman, R. S.; Cabral, J.; Hatton, T.A. 1987, in preparation.
11. Goklen, K. E. Ph.D. Thesis, M.I.T., Cambridge MA, 1986.
12. Sheu, E.; Goklen, K. E.; Hatton, T.A.; Chen, S.-H. Biotech. Prog. 1986, 2, 175-86.
13. Hinze, W. L., personal communication.
14. Hiemenz, P. C. "Principles of Colloid and Surface Chemistry"; Marcel Dekker: New York, 1977; p.369.
15. Meier, P.; Imre, E.; Fleschar, M.; Luisi, P. L. In "Surfactants in Solution"; Mittal, K. L.; Lindman, B., Eds.; Plenum Press: New York, 1984.
16. Woll, J.; Hatton, T. A., work in progress.
17. Rawn, J. D. "Biochemistry"; Harper and Row: New York, 1983.
18. Thien, M. P.; Lee, K. K.; Thompson, K.; Hatton, T. A. 1987, in preparation.
19. Bull, H. B.; Breese, K. Acta Biochem and Biophysics 1974, 161, 665-70.
20. Eriksen, J. C. Acta Chem. 1963, 17.
21. Murkherjee, P.; Cardinal, J. R.; Desai, N. R. In "Micellization, Solubilization and Microemulsions"; Mittal, K. L., Ed.; Plenum Press: New York, 1976.
22. Wong, M.; Thomas, J. K.; Gratzel, M. JACS 1976, 98, 2391-7.
23. Tanford, C. R. "The Hydrophobic Effect"; Wiley: New York, 1980.

RECEIVED May 11, 1987

Chapter 10

Equilibrium Solubilization of Benzene in Micellar Systems and Micellar-Enhanced Ultrafiltration of Aqueous Solutions of Benzene

George A. Smith[1,2], Sherril D. Christian[1,2], Edwin E. Tucker[1,2], and John F. Scamehorn[2,3]

[1]Department of Chemistry, University of Oklahoma, Norman, OK 73019
[2]Institute for Applied Surfactant Research, University of Oklahoma, Norman, OK 73019
[3]School of Chemical Engineering and Materials Science, University of Oklahoma, Norman, OK 73019

An automated vapor pressure method has been used to obtain highly precise values of the partial pressure of benzene as a function of concentration in aqueous solutions of sodium dodecylsulfate (at 15 to 45°C) and 1-hexadecylpyridinium chloride (at 25 to 45°C). Solubilization isotherms and the dependence of benzene activity on the intramicellar composition are inferred from the measurements and related to probable micellar structures and changes in structure accompanying the solubilization of benzene. Calculations are made to determine the efficiency of micellar-enhanced ultrafiltration (MEUF) as a process for purifying water streams contaminated by benzene.

The mobility of solute species in aqueous media and the transfer of these solutes to other phases can be greatly influenced by their association with ordered entities such as surfactant micelles. Thus, the effectiveness of micellar-enhanced ultrafiltration (MEUF) in removing organic (1-4) and metal ion (5, 6) contaminants from aqueous streams owes to the fact that surfactant micelles containing these contaminants are too large to pass through the pores of an ultrafilter. In several column chromatographic methods, separations are achieved because micellar moieties, in moving or fixed phases, are able to diminish the concentration of free organic molecules in contiguous bulk phases (7-11). The ability of aqueous micellar solutions to dissolve molecules that would otherwise be practically insoluble in water can also serve as the basis for separating compounds that are very similar in most molecular properties (12, 13).

Considering the importance of micellar aggregates in separations, it is unfortunate that our knowledge of solute-micelle equilibria is quite limited, both as regards the dependence of solute activities on the intramicellar mole fractions of surfactant and organic compound, and in relation to the influence of total surfactant concentration and temperature on solubilization behavior. Only rarely have measurements been obtained with sufficient accuracy to permit tests to be made of theories of solubilization.

The lack of solubilization data impinges directly on our ability to design procedures for removing contaminants from aqueous streams by micellar-enhanced ultrafiltration. In MEUF, an aqueous stream containing a dissolved contaminant plus added surfactant is passed through an ultrafilter. In several studies (1-4) it has been shown that the permeate stream has a concentration of organic solute approximately equal to that of the unsolubilized or monomeric organic species in the retentate stream. As a result, MEUF can be an extremely effective technique for cleaning contaminated water, producing a purified stream with very small concentrations of organic solutes.

In our laboratories, extensive use has been made of vapor pressure (14-18) and membrane methods (2, 3, 19, 20) to infer thermodynamic results for ternary aqueous systems containing an ionic or a nonionic surfactant and an organic solute. The most precise solubilization measurements ever reported have been obtained with an automated vapor pressure apparatus for volatile hydrocarbon solutes such as cyclohexane and benzene, dissolved in aqueous solutions of sodium octylsulfate and other ionic surfactants (15, 16). A manual vapor pressure apparatus has been employed to obtain somewhat less precise results for solutes of lower volatility (17, 18). Recently, semi-equilibrium dialysis (19, 20) and MEUF (2) methods have been used to investigate solute-surfactant systems in which the organic solubilizates are too involatile to study by ordinary vapor pressure methods.

The present report describes new results for benzene at temperatures in the range 15 to 45°C, solubilized in aqueous solutions of sodium dodecylsulfate (SDS) and 1-hexadecylpyridinium chloride (referred to as cetylpyridinium chloride or CPC). The solute activity vs. concentration data provide insight into the nature of chemical and structural effects responsible for the solubilization of benzene by aqueous micellar systems; in addition, the results find direct use in predicting the performance of MEUF in removing dissolved benzene from aqueous streams.

Experimental

The solubilization results reported here were obtained with an automated vapor pressure apparatus described previously (21). Benzene samples, from an external reservoir at 50°C, were added incrementally to the main measuring system by means of a 6-port HPLC valve. Successive increments of benzene are allowed to flash from the valve into the main solution reservoir; these samples contain 2.907×10^{-4} moles of benzene, with a reproducibility better than 1 part in 6,000 or 7,000 (22).

The benzene used was Analytical Reagent Grade from Mallinckrodt
Chemical Company, distilled through a 25-plate bubble-cap column
and stored in vapor contact with desiccant prior to use.
High-quality 1-hexadecylpyridinium chloride (CPC) from Hexcel
Corporation was used without further purification. Sodium
dodecylsulfate (SDS) was HPLC-grade chemical from Fisher
Scientific Company, purified by recrystallization from an ethan-
ol-water mixture.

Results and Discussion

Table I lists experimental results, comprising derived values of
the fugacity of benzene at known total molarity in the aqueous
phase, [B], and known molarity of 1-hexadecylpyridinium chloride
[CPC] or sodium dodecylsulfate [SDS]. Fugacities have been
calculated from total pressures by subtracting the vapor pressure
of the aqueous solution in the absence of benzene from the
measured total pressure and correcting for the small extent of
nonideality of the vapor phase (15, 22). Results are given for
temperatures varying from 25 to 45°C for the CPC systems and 15 to
45°C for the SDS systems.

The data in Table I may be used to infer values of K, the
solubilization equilibrium constant or partition coefficient
defined by

$$K = X_B / c_B$$

where X_B is the mole fraction of benzene in the intramicellar
"solution" (20) and c_B is the concentration of benzene in
monomeric form in the bulk aqueous solution. In the case of CPC,
the surfactant molecules are assumed to exist entirely in micellar
form (23), although in calculating K values for the SDS systems,
small corrections are made to account for the concentration of the
surfactant that is not in micelles (20, 24). It is assumed that
the concentration of monomeric organic solute can be calculated
from the fugacity of the organic solute (practically equal to the
partial pressure), using the Henry's law constant inferred from
data for the solute dissolved in pure water, with a small
correction for "salting-out" by the ionic surfactant solution
(15, 16).

The primary results in Table I may also be processed to yield
values of the benzene activity coefficient in the intramicellar
solution, γ_B, defined as $f_B/(f_B^o X_B)$, where f_B is the fugacity of
benzene in equilibrium with the aqueous surfactant solution, and
f_B^o is the fugacity of pure benzene at the given temperature.
Figures 1-4 are plots of the solubilization constant (K) and the
benzene activity coefficient (γ_B) against the intramicellar mole
fraction of benzene (X_B) for the surfactants CPC and SDS at the
indicated temperatures.

Table I. Benzene fugacities above aqueous solutions containing known concentrations of benzene and 1-hexadecylpyridinium chloride (CPC) at temperatures varying from 25 to 45° C

Benzene into CPC at 25° C.

f (Torr)	[Benzene]	[CPC]
2.1023E+00	2.5710E-03	1.0626E-01
4.1671E+00	5.1480E-03	1.0624E-01
6.2088E+00	7.7281E-03	1.0621E-01
8.2292E+00	1.0311E-02	1.0619E-01
1.0230E+01	1.2896E-02	1.0616E-01
1.1208E+01	1.5484E-02	1.0614E-01
1.4168E+01	1.8074E-02	1.0611E-01
1.6104E+01	2.0668E-02	1.0609E-01
1.8020E+01	2.3264E-02	1.0607E-01
1.9911E+01	2.5863E-02	1.0604E-01
2.1786E+01	2.8464E-02	1.0602E-01
2.3637E+01	3.1069E-02	1.0599E-01
2.5467E+01	3.3676E-02	1.0597E-01
2.7268E+01	3.6287E-02	1.0594E-01
2.9054E+01	3.8901E-02	1.0592E-01
3.0809E+01	4.1518E-02	1.0589E-01
3.2546E+01	4.4138E-02	1.0587E-01
3.4256E+01	4.6763E-02	1.0584E-01
3.5937E+01	4.9391E-02	1.0582E-01
3.7594E+01	5.2023E-02	1.0579E-01
3.9227E+01	5.4658E-02	1.0577E-01
4.0832E+01	5.7298E-02	1.0574E-01
4.2412E+01	5.9942E-02	1.0572E-01
4.3960E+01	6.2590E-02	1.0569E-01
4.5483E+01	6.5242E-02	1.0567E-01
4.6974E+01	6.7899E-02	1.0564E-01
4.8443E+01	7.0560E-02	1.0562E-01
4.9880E+01	7.3226E-02	1.0559E-01
5.1285E+01	7.5896E-02	1.0557E-01
5.2662E+01	7.8571E-02	1.0554E-01
5.4013E+01	8.1251E-02	1.0552E-01
5.5334E+01	8.3934E-02	1.0549E-01
5.6627E+01	8.6623E-02	1.0547E-01
5.7894E+01	8.9315E-02	1.0544E-01
5.9127E+01	9.2013E-02	1.0542E-01
6.0331E+01	9.4715E-02	1.0539E-01
6.1505E+01	9.7423E-02	1.0536E-01
6.2651E+01	1.0013E-01	1.0534E-01
6.3764E+01	1.0285E-01	1.0531E-01
6.4847E+01	1.0557E-01	1.0529E-01
6.5908E+01	1.0830E-01	1.0526E-01

Benzene into CPC at 35° C.

f (Torr)	[Benzene]	[CPC]
3.0480E+00	2.6636E-03	1.1897E-01
6.0576E+00	5.3335E-03	1.1894E-01
9.0289E+00	8.0096E-03	1.1891E-01
1.1973E+01	1.0689E-02	1.1888E-01
1.4877E+01	1.3376E-02	1.1885E-01
1.7770E+01	1.6063E-02	1.1883E-01
2.0618E+01	1.8757E-02	1.1880E-01
2.3448E+01	2.1453E-02	1.1877E-01
2.6256E+01	2.4152E-02	1.1874E-01
2.9033E+01	2.6856E-02	1.1871E-01
3.1776E+01	2.9565E-02	1.1868E-01
3.4495E+01	3.2277E-02	1.1865E-01
3.7186E+01	3.4994E-02	1.1862E-01
3.9839E+01	3.7717E-02	1.1859E-01
4.2465E+01	4.0444E-02	1.1857E-01
4.5058E+01	4.3176E-02	1.1854E-01
4.7609E+01	4.5916E-02	1.1851E-01
5.0125E+01	4.8662E-02	1.1848E-01
5.2617E+01	5.1412E-02	1.1845E-01
5.5076E+01	5.4166E-02	1.1842E-01
5.7494E+01	5.6929E-02	1.1839E-01
5.9883E+01	5.9696E-02	1.1836E-01
6.2236E+01	6.2469E-02	1.1833E-01
6.4550E+01	6.5249E-02	1.1830E-01
6.6817E+01	6.8038E-02	1.1827E-01
6.9052E+01	7.0832E-02	1.1824E-01
7.1238E+01	7.3637E-02	1.1821E-01
7.3384E+01	7.6448E-02	1.1818E-01
7.5504E+01	7.9263E-02	1.1815E-01
7.7546E+01	8.2095E-02	1.1812E-01
7.9581E+01	8.4927E-02	1.1809E-01
8.1582E+01	8.7765E-02	1.1806E-01
8.3550E+01	9.0608E-02	1.1803E-01
8.5466E+01	9.3462E-02	1.1800E-01
8.7351E+01	9.6321E-02	1.1797E-01
8.9198E+01	9.9187E-02	1.1794E-01
9.1006E+01	1.0206E-01	1.1791E-01
9.2773E+01	1.0494E-01	1.1788E-01
9.4504E+01	1.0783E-01	1.1784E-01
9.6179E+01	1.1073E-01	1.1781E-01

Benzene into CPC at 45° C.

f (Torr)	[Benzene]	[CPC]
4.2112E+00	3.0285E-03	1.5133E-01
8.3790E+00	6.0650E-03	1.5129E-01
1.2504E+01	9.1093E-03	1.5125E-01
1.6576E+01	1.2165E-02	1.5121E-01
2.0619E+01	1.5224E-02	1.5117E-01
2.4618E+01	1.8292E-02	1.5112E-01
2.8575E+01	2.1368E-02	1.5108E-01
3.2499E+01	2.4449E-02	1.5104E-01
3.6383E+01	2.7537E-02	1.5100E-01
4.0219E+01	3.0636E-02	1.5095E-01
4.4021E+01	3.3740E-02	1.5091E-01
4.7781E+01	3.6853E-02	1.5087E-01
5.1509E+01	3.9971E-02	1.5082E-01
5.5196E+01	4.3097E-02	1.5078E-01
5.8842E+01	4.6231E-02	1.5074E-01
6.2449E+01	4.9374E-02	1.5070E-01
6.6044E+01	5.2515E-02	1.5065E-01
6.9555E+01	5.5678E-02	1.5061E-01
7.3030E+01	5.8848E-02	1.5057E-01
7.6499E+01	6.2016E-02	1.5052E-01
7.9914E+01	6.5196E-02	1.5048E-01
8.3246E+01	6.8398E-02	1.5043E-01
8.6527E+01	7.1612E-02	1.5039E-01
8.9831E+01	7.4815E-02	1.5035E-01
9.3054E+01	7.8039E-02	1.5030E-01
9.6184E+01	8.1289E-02	1.5026E-01
9.9366E+01	8.4518E-02	1.5021E-01
1.0244E+02	8.7775E-02	1.5017E-01
1.0544E+02	9.1056E-02	1.5012E-01
1.0839E+02	9.4346E-02	1.5008E-01
1.1131E+02	9.7642E-02	1.5003E-01
1.1424E+02	1.0093E-01	1.4999E-01
1.1707E+02	1.0425E-01	1.4994E-01
1.1982E+02	1.0759E-01	1.4989E-01
1.2262E+02	1.1091E-01	1.4985E-01
1.2534E+02	1.1425E-01	1.4980E-01
1.2794E+02	1.1762E-01	1.4976E-01
1.3051E+02	1.2101E-01	1.4971E-01
1.3315E+02	1.2436E-01	1.4966E-01

a All concentrations in mol-l^{-1}

Continued on next page.

Table I Continued.   Benzene fugacities above aqueous
solutions containing known concentrations of benzene and
sodium dodecylsulfate (SDS) at temperatures varying from
15 to 45° C [a]

Benzene into SDS at 15° C.

f (Torr)	[Benzene]	[SDS]
2.2219E+00	2.5013E-03	1.0044E-01
4.4041E+00	5.0092E-03	1.0041E-01
6.5598E+00	7.5209E-03	1.0039E-01
8.6790E+00	1.0039E-02	1.0037E-01
1.0757E+01	1.2563E-02	1.0035E-01
1.2797E+01	1.5094E-02	1.0032E-01
1.4796E+01	1.7633E-02	1.0030E-01
1.6754E+01	2.0177E-02	1.0028E-01
1.8668E+01	2.2730E-02	1.0026E-01
2.0544E+01	2.5289E-02	1.0023E-01
2.2380E+01	2.7855E-02	1.0021E-01
2.4168E+01	3.0430E-02	1.0019E-01
2.5917E+01	3.3011E-02	1.0017E-01
2.7610E+01	3.5603E-02	1.0014E-01
2.9263E+01	3.8202E-02	1.0012E-01
3.0866E+01	4.0809E-02	1.0010E-01
3.2416E+01	4.3427E-02	1.0007E-01
3.3922E+01	4.6052E-02	1.0005E-01
3.5373E+01	4.8688E-02	1.0003E-01
3.6773E+01	5.1332E-02	1.0000E-01
3.8126E+01	5.3986E-02	9.9981E-02
3.9435E+01	5.6647E-02	9.9957E-02
4.0686E+01	5.9319E-02	9.9934E-02
4.1895E+01	6.1999E-02	9.9910E-02
4.3060E+01	6.4687E-02	9.9886E-02

Benzene into SDS at 25° C.

f (Torr)	[Benzene]	[SDS]
3.2476E+00	2.3854E-03	1.0407E-01
6.4517E+00	4.7776E-03	1.0404E-01
9.6103E+00	7.1771E-03	1.0402E-01
1.2724E+01	9.5839E-03	1.0400E-01
1.5788E+01	1.1999E-02	1.0398E-01
1.8805E+01	1.4422E-02	1.0395E-01
2.1776E+01	1.6852E-02	1.0393E-01
2.4693E+01	1.9292E-02	1.0391E-01
2.7562E+01	2.1740E-02	1.0389E-01
3.0374E+01	2.4199E-02	1.0386E-01
3.3134E+01	2.6666E-02	1.0384E-01
3.5840E+01	2.9143E-02	1.0382E-01
3.8490E+01	3.1629E-02	1.0379E-01
4.1084E+01	3.4127E-02	1.0377E-01
4.3616E+01	3.6635E-02	1.0375E-01
4.6091E+01	3.9154E-02	1.0373E-01
4.8509E+01	4.1683E-02	1.0370E-01
5.0862E+01	4.4225E-02	1.0368E-01
5.3146E+01	4.6779E-02	1.0365E-01
5.5364E+01	4.9346E-02	1.0363E-01
5.7519E+01	5.1925E-02	1.0361E-01
5.9601E+01	5.4519E-02	1.0358E-01
6.1622E+01	5.7123E-02	1.0356E-01
6.3574E+01	5.9742E-02	1.0353E-01
6.5460E+01	6.2372E-02	1.0351E-01

Benzene into SDS at 35° C.

f (Torr)	[Benzene]	[SDS]
4.4818E+00	2.2401E-03	1.0876E-01
8.9074E+00	4.4889E-03	1.0874E-01
1.3276E+01	6.7466E-03	1.0872E-01
1.7586E+01	9.0136E-03	1.0869E-01
2.1840E+01	1.1289E-02	1.0867E-01
2.6058E+01	1.3570E-02	1.0865E-01
3.0206E+01	1.5862E-02	1.0863E-01
3.4308E+01	1.8162E-02	1.0860E-01
3.8333E+01	2.0475E-02	1.0858E-01
4.2299E+01	2.2798E-02	1.0856E-01
4.6209E+01	2.5130E-02	1.0854E-01
5.0059E+01	2.7473E-02	1.0851E-01
5.3862E+01	2.9823E-02	1.0849E-01
5.7572E+01	3.2191E-02	1.0847E-01
6.1225E+01	3.4569E-02	1.0844E-01
6.4815E+01	3.6959E-02	1.0842E-01
6.8336E+01	3.9361E-02	1.0840E-01
7.1781E+01	4.1777E-02	1.0837E-01
7.5162E+01	4.4206E-02	1.0835E-01
7.8460E+01	4.6650E-02	1.0833E-01
8.1691E+01	4.9107E-02	1.0830E-01
8.4840E+01	5.1580E-02	1.0828E-01
8.7919E+01	5.4066E-02	1.0825E-01
9.0918E+01	5.6569E-02	1.0823E-01
9.3839E+01	5.9086E-02	1.0820E-01
9.6673E+01	6.1622E-02	1.0818E-01

Benzene into SDS at 45° C.

f (Torr)	[Benzene]	[SDS]
5.8397E+00	2.0377E-03	1.1220E-01
1.1588E+01	4.0900E-03	1.1218E-01
1.7274E+01	6.1509E-03	1.1216E-01
2.2911E+01	8.2174E-03	1.1214E-01
2.8496E+01	1.0290E-02	1.1212E-01
3.4014E+01	1.2373E-02	1.1210E-01
3.9476E+01	1.4464E-02	1.1208E-01
4.4923E+01	1.6552E-02	1.1205E-01
5.0324E+01	1.8647E-02	1.1203E-01
5.5576E+01	2.0770E-02	1.1201E-01
6.0784E+01	2.2898E-02	1.1199E-01
6.5967E+01	2.5028E-02	1.1197E-01
7.1049E+01	2.7176E-02	1.1195E-01
7.6051E+01	2.9338E-02	1.1192E-01
8.1003E+01	3.1507E-02	1.1190E-01
8.5938E+01	3.3676E-02	1.1188E-01
9.0770E+01	3.5864E-02	1.1186E-01
9.5466E+01	3.8079E-02	1.1183E-01
1.0014E+02	4.0296E-02	1.1181E-01
1.0473E+02	4.2528E-02	1.1179E-01
1.0933E+02	4.4755E-02	1.1177E-01
1.1379E+02	4.7010E-02	1.1174E-01
1.1812E+02	4.9290E-02	1.1172E-01
1.2241E+02	5.1577E-02	1.1170E-01

[a]
All concentrations in mol-l^{-1}

Table I and Figures 1-4 contain a wealth of information about the solubilization of benzene in aqueous surfactant micelles.   Plots of K vs. X_B exhibit shallow minima in the case of the SDS solutions, and rather more pronounced minima for the CPC solutions.   The plots of γ_B vs. X_B show corresponding maxima, reflecting the fact that K and γ_B are related reciprocally by $K = 1/(\gamma_B c_B^{\,o})$, where $c_B^{\,o}$ is the monomer concentration of benzene in the aqueous phase at saturation. (The minimum in K and the maximum in γ_B for the CPC solutions, shown in Figure 1, are not quite reached at the benzene concentrations attainable with the automated vapor pressure apparatus. The automated apparatus is restricted to operating at partial pressures less than about 70% of the vapor pressure of pure liquid benzene. However, the manual apparatus can be used for measurements almost to saturation, and results obtained with this apparatus show extrema in K and γ_B at approximately X = 0.55.)

It seems likely that the cationic CPC micelles, which have a large positive charge at or near the micellar surface, interact attractively with the π-molecular orbital system of benzene, and that this interaction contributes to the fact that the solubilization constant for benzene in CPC is approximately twice as large as that in SDS micelles. A preferential interaction between cationic surfactants and aromatic solutes has been reported by several groups of investigators (25-27), and recent work in our laboratory shows that 1-hexadecyltrimethylammonium bromide micelles also solubilize benzene more effectively than do the anionic alkylsulfate surfactant micelles (28).   Thus, the tendency of benzene molecules to solubilize near the surface of the cationic micelles, at low X_B values, may lead to a partial saturation of surface "sites" by benzene, diminishing the ability of additional benzene molecules to bind near the surface. Such an effect could be responsible for the initial increase in activity coefficient that occurs, particularly in the CPC solutions, as X_B increases.

Another effect which probably contributes to the increase in γ_B that occurs as benzene is added to ionic surfactant solutions is the decrease in micellar surface charge that is caused by inserting benzene molecules into either cationic or anionic micelles. Diminishing the surface charge should significantly decrease the importance of the ion-induced dipole effect that is partly responsible for solubilizing benzene (17).   Finally, it seems necessary to conclude that benzene molecules will tend to solubilize favorably within the hydrocarbon core region of the micelles at any intramicellar composition. Perhaps most important in supporting this conclusion is the observation that only small changes in K or γ_B occur in the benzene-CPC and benzene-SDS systems throughout wide ranges of intramicellar composition.

The ultimate decrease in benzene activity coefficients at the largest X_B values may owe to several factors, including a possible diminution of the so-called Laplace pressure (29, 30, 14),

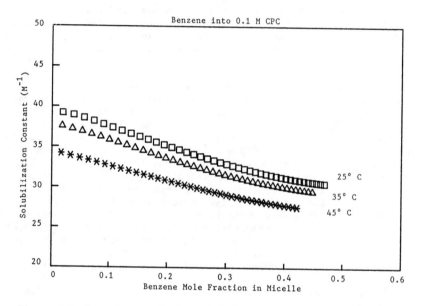

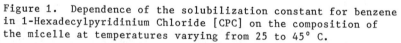

Figure 1. Dependence of the solubilization constant for benzene
in 1-Hexadecylpyridinium Chloride [CPC] on the composition of
the micelle at temperatures varying from 25 to 45° C.

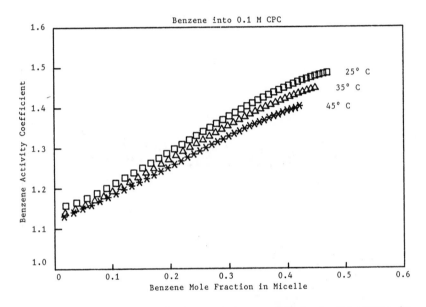

Figure 2. Dependence of the activity coefficient of benzene in 1-Hexadecylpyridinium Chloride [CPC] on the composition of the micelle at temperatures varying from 25 to 45° C.

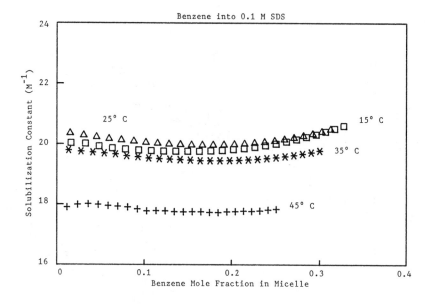

Figure 3. Dependence of the solubilization constant for benzene
in Sodium Dodecylsulfate [SDS] on the composition of the micelle
at temperatures varying from 15 to 45° C.

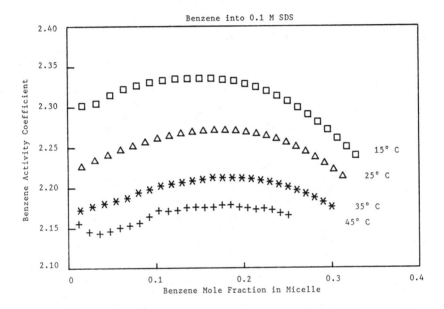

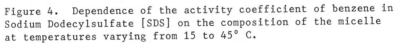

Figure 4.   Dependence of the activity coefficient of benzene in Sodium Dodecylsulfate [SDS] on the composition of the micelle at temperatures varying from 15 to 45° C.

reflecting the reduction in curvature of the micellar surface that probably occurs as more benzene is incorporated in the micelle. However, the tendency of γ_B to approach unity as X_B increases to large values may simply reflect the fact that the micellar interior must more and more closely resemble liquid benzene as the mole fraction of benzene becomes greater; any reasonable theory of concentrated solutions should predict this effect.

The temperature dependence of K and γ_B, shown for the CPC and SDS systems in Figures 1-4, illustrates several important features that are typical of the solubilization of hydrocarbons by aqueous micellar solutions. The K values attain a maximum for each system in the vicinity of 25°C, indicating that ΔH for solubilizing benzene reaches a value of zero in this range; a similar effect has been observed previously for benzene in sodium octylsulfate micelles (15, 16). The thermodynamic constants for the solubilization of benzene by micelles closely resemble results for the hydrophobic association of hydrocarbon molecules and moieties in aqueous solution (22, 31-34). The very large negative heat capacity change accompanying the solubilization of benzene is thought to indicate that increasing the temperature diminishes the extent of the ordered-water region surrounding hydrocarbon molecules in aqueous solution (34).

The thermodynamic quantities derived from the temperature dependence of the activity coefficients (γ_B) are not so difficult to interpret. Neither the enthalpy nor the heat capacity changes for transferring benzene from the pure liquid phase into the micellar interior exhibit the anomalies that are characteristic of the transfer of a hydrocarbon molecule to or from the dilute aqueous solution phase. The relatively small decrease in activity coefficient that occurs as the temperature increases indicates that the transfer of benzene from the pure liquid phase into the micelles is slightly endothermic throughout the range of X_B included in the experiments. The change in partial molar enthalpy of benzene for the transfer, near the midpoint of the temperature and X_B range, is approximately 400 cal/mole for the SDS solutions and 350 cal/mole for the CPC solutions. The excess entropy changes for the transfer are surprisingly small, having values of approximately -0.3 cal $mol^{-1}K^{-1}$ for the SDS micelles and +0.5 cal $mol^{-1} K^{-1}$ for the CPC micelles. Taken together, the thermodynamic results do not support the concept that the transfer of benzene into either the SDS or the CPC micelles involves strong localized adsorption at specific sites within the micelle or near the micelle surface.

In utilizing the solubilization measurements to estimate the ability of MEUF to remove dissolved benzene from water, we arbitrarily assume that the feed solution contains 50 mM surfactant and 1 mM benzene. The solution is ultrafiltered until 80% of the volume of the solution is removed as permeate; if the water were to be recycled to the plant, this would correspond to a recycle ratio of 80%. For these assumed conditions, the

Table II: Performance of MEUF in Removal of Benzene from Water[a]

Surfactant	Temperature (°C)	Final Concentration in Permeate (mM)	Final Concentration in Retentate (mM)	Rejection (%)
CPC[b]	25	0.360	3.56	89.89
CPC	35	0.371	3.52	89.45
CPC	45	0.397	3.41	88.36
SDS[c]	15	0.555	2.78	80.04
SDS	25	0.549	2.80	80.41
SDS	35	0.559	2.77	79.79
SDS	45	0.601	2.60	76.80

[a] Feed: [benzene] = 1 mM; [surfactant] = 50 mM.
Permeate/feed = 0.8.

[b] CPC = 1-hexadecylpyridinium chloride

[c] SDS = sodium dodecylsulfate

concentrations of benzene in the permeate and retentate streams from the process are listed in Table II. The rejection, a parameter commonly used in membrane science, is also given in Table II. The rejection is defined as

rejection(%) = 100{1- [benzene in permeate]/[benzene in retentate]}

Under the conditions specified in the table, when CPC is the surfactant, the concentration of benzene in the permeate is 10 times smaller than that in the retentate product from the process and less than 40% of that in the feed. With SDS as the surfactant, the permeate has a factor of 5 smaller concentration of benzene than the retentate and is less than 60% as concentrated as the feed. Thus the preferential interaction of the cationic surfactant, CPC, with benzene makes CPC superior to SDS in removing benzene from water.

The data in Table II indicate that changing the temperature only slightly affects the purity of the permeate stream, the maximum difference being 7% between 25 and 45°C for benzene/SDS. This result is important practically in showing that MEUF performance in a real process will not be substantially affected by temperature variations in field waters. Stability of the process with respect to variations in temperature is a positive feature in industrial separations.

The present experimental results show that the relative tendency of benzene to solubilize in the surfactant micelles decreases slightly as the benzene concentration increases from near zero to higher values. Thus, the MEUF separation becomes slightly poorer with increased loading of benzene, an effect also observed for phenol and the cresols (20, 28). At still higher benzene concentrations, the separation will improve, because the solubilization constant eventually reaches a minimum and increases as X_B increases. These considerations show that in predicting MEUF performance, it is necessary to have accurate solubilization results for contaminants throughout wide ranges of X_B.

Although the rejections shown in Table II (75-90%) are not very good for industrial applications, we have shown that rejections as great as 99.8% can be obtained for the removal of other organic solutes (2, 6). Thus, MEUF is a promising industrial separation, but benzene may not be an optimum candidate for removal using MEUF.

Acknowledgments

The research reported here has been supported by the National Science Foundation (Grant CHE-8402866) and the Department of Energy (Contract DE-AS05-84ER3175).

Literature Cited

1. Leung, P. S. In "Ultrafiltration Membranes and Applications"; Cooper, A. R., Ed.; Plenum: New York, 1979; p. 415.
2. Dunn, R. O.; Scamehorn, J. F.; Christian, S. D. Sep. Sci. Technol. 1985, 20, 257.
3. Dunn, R. O.; Scamehorn, J. F.; Christian, S. D. Sep. Sci. Technol. In Press.
4. Gibbs, L. L.; Scamehorn, J. F.; Christian, S. D. J. Mem. Membrane Sci. In Press.
5. Scamehorn, J. F.; Ellington, R. T.; Christian, S. D.; Penney, B. W.; Dunn, R. O.; Bhat, S. R. A. I. Ch. E. Symp. Series, In Press.
6. Scamehorn, J. F.; Harwell, J. H. In "Surfactants and Chemical Engineering"; Wasan, D. T.; Shah, D. O.; Ginn, M. E., Ed.; Marcel Dekker: New York, In Press.
7. Armstrong, D. W. Separations and Purification Methods 1985, 14, 213.
8. Armstrong, D. W.; Fendler, J. H. Biochim. Biophys. Acta 1977, 75, 418.
9. Maley, F.; Guarino, D. U. Biochim. Biophys. Res. Comm. 1977, 77, 1425.
10. Graham, J. A.; Rogers, L. B.; J. Chromatogr. Sci. 1980, 18, 614
11. Armstrong, D. W.; Nome, F. Anal. Chem. 1981, 53, 1662.
12. Borgerding, M. F.; Hinze, W. L. Anal. Chem. 1985, 57, 2183.
13. Janini, G. M.; Attari, S. A. Anal. Chem. 1983, 55, 659.
14. Christian, S. D.; Tucker, E. E.; Lane, E. H. J. Colloid Interface Science 1981, 84, 423.
15. Tucker, E. E.; Christian, S. D. Faraday Symp. Chem. Soc. 1982, 17, 11.
16. Tucker, E. E.; Christian, S. D. J. Colloid Interface Science 1985, 104, 562.
17. Christian, S. D.; Tucker, E. E.; Smith, G. A.; Bushong, D. S. J. Colloid Interface Science 1982, 89, 514.
18. Christian, S. D.; Smith, L. S.; Bushong, D. S.; Tucker, E. E. J. Colloid Interface Science 1982, 89, 514.
19. Christian, S. D.; Smith, G. A.; Tucker, E. E.; Scamehorn, J. F. Langmuir 1985, 1, 564.
20. Smith, G. A.; Christian, S. D.; Tucker, E. E.; Scamehorn, J. F. J. Solution Chem. 1986, 15, 519.
21. Tucker, E. E.; Christian, S. D. J. Chem. Thermodyn. 1979, 11, 1137.
22. Tucker, E. E.; Christian, S. D. J. Solution Chem. 1981, 10, 1.
23. Bushong, D. S.; Ph. D. Dissertation, University of Oklahoma, Norman, Oklahoma, 1985.
24. Abu-Hamdiyyah, M.; Mysels, K. J. J. Phys. Chem. 1967, 71, 418.
25. Nagarajan, R.; Chaiko, M. A.; Ruckenstein, E. J. Phys. Chem. 1984, 88, 2916.
26. Rehfeld, S. J.; J. Phys. Chem. 1971, 75, 3905.

27. Hirose, C.; Sepulveda, L. J. Phys. Chem 1981, 85, 3689.
28. Smith, G. A. Ph. D. Dissertation, University of Oklahoma, Norman, OK, In Preparation.
29. Mukerjee, P. In "Surface Chemistry of Surfactants"; K. L. Mittal, Ed.; Plenum Press, New York, 1978, p. 153.
30. Matheson, I. B. C.; King, Jr., A. D. J. Colloid Interface Sci. 1978, 66, 464.
31. Frank, H. S.; Evans, M. W. J. Chem. Phys. 1945, 13, 507.
32. Kauzmann, W. Adv. Protein Chem. 1959, 14, 1.
33. Franks, F. In "Water: A Comprehensive Treatise"; Vol. 4, Chapter 1, Plenum Press, New York, 1975.
34. Christian, S. D.; Tucker, E. E. J. Solution Chem. 1982, 11, 749.

RECEIVED October 27, 1986

CYCLODEXTRINS

Chapter 11

Cyclodextrin Use in Separations

J. Szejtli[1], B. Zsadon[2], and T. Cserhati[3]

[1]Chinoin Pharmaceutical-Chemical Works H-1026 Budapest Endrodi S. 38/40, Hungary
[2]Institute of Chemical Technology, Eotvos Lorand University, H-1088, Muzeum krt. 6-8, Hungary
[3]Research Institute for Plant Protection, H-1020, Budapest, Hermann O. 15, Hungary

The central cavity of the cylinder-shaped cyclodextrins behaves as an empty capsule: it can accommodate so-called guest molecules of appropriate size, shape, and polarity. This "molecular encapsulation" can be utilized for stabilization and for enhancement of solubility of drugs, vitamins, flavors, etc., and utilizing the selectivity of the inclusion complexation, it can be applied for separation of substances, either by non-chromatographic methods, or chromatographic methods.

The products of partial enzymic or acidic degradation of starch are called dextrins, which are heterogeneous, amorphous, hygroscopic, sticky substances. There is, however, a starch degrading enzyme, which produces 3 crystallinze, homogeneous, non-hygroscopic cyclodextrins of different molecular size (1) (Figure 1).
      These cyclodextrins are cylinder-shaped molecules with an axial void cavity. Their outer surface is hydrophilic, therefore the cyclodextrins are soluble in water. Their cavity, however, is of apolar character (Figure 2). As a consequence of this structure, cyclodextrins can include other apolar molecules of appropriate dimensions and bind them through apolar-apolar interactions (2) (Figure 3). These inclusion complexes are crystalline substances. As a result of such a "molecular encapsulation", characteristic properties of the included substances will be changed (3,4). Volatile or gaseous substances can be converted into stable, crystalline substances, oxygen sensitive materials are protected against atmospheric oxidation, solubility of poorly soluble substances is improved, bioavailability of scarcely soluble drugs is enhanced, volatile, sensitive flavours and fragrances can be stored without loss, marketable drugs or pesticides can be prepared from compounds of intolerable odors, enzymatic reactions of lipophilic substances can be accelerated, complexing toxic or

0097-6156/87/0342-0200$06.00/0

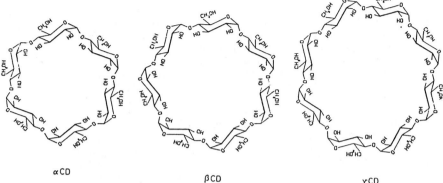

αCD

βCD

γCD

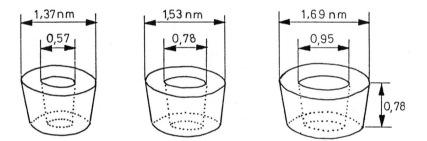

Figure 1. Chemical structure and molecular dimensions of α-, β- and γ-cyclodextrins.

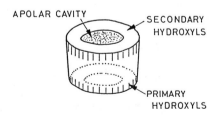

Figure 2. Schematic representation of the cyclodextrin "cylinder".

inhibitory substances microorganism can be protected and stimulated
to higher metabolic activity, selectivity and rate of chemical
reactions can be improved, etc. The number of possible applications
seems to be inexhaustible.

Substitution of one or more hydroxyls of a cyclodextrin in most
cases results in better water soluble derivatives (5). Additionally
such cyclodextrin derivatives can exhibit modified complex forming
capacity. Cyclodextrin can be polymerized by appropriate bi- or
polyfunctional agents to oligomers, long-chain polymers or to crosslinked networks or can be immobilized on various supports. The low-
molecular cyclodextrin-oligomers are readily soluble in water. The
polymers (mol. weight over 10,000) are swelling gels (3). The
swelling polymers can be prepared in bead form (6) (Fig.4.).

The rigid structure of the cyclodextrin host results in well
defined but different inclusion and interaction patterns for any
potential guest molecule. Treating a mixture of compounds with a
dissolved or solid, immobilized CD, leads to the formation of
inclusion complexes of different stability and solubility. Consequently separations can be based either on strongly modified solubility in water of the CD-complex of a certain component, or on the
difference of their K_{diss} values.

Non-chromatographic separation by cyclodextrins

A characteristic feature of non-chromatographic separations utilizing
cyclodextrins is that they are aimed at preparative separations.
Unfortunately only incomplete separations or enrichments can be
attained. By repeating the separations in multistage processes, the
required component can be enriched on preparative, and even industrial, scale. Many examples have been published both for partial
separation of compounds, isomers, or enantiomers through selective
crystallization of their complexes (3).

Upon incorporation of cyclodextrins in membranes, or dissolution of cyclodextrins in one or the membrane-separated liquid phases,
the permeation rate of the complexed guest-molecule can be modified
considerably (7-9). This offers another possibility for the enrichment of the selected component, nevertheless no rapid quantitative
separation can be attained. Therefore separations which satisfy the
requirements of the separation scientist can be achieved only by
chromatographic methods.

Chromatographic applications of cyclodextrins

Cyclodextrins and their derivatives can be applied in all current
types of chromatography (10-12). Table I illustrates the actual
known possibilities which does not implicate that no further combinations (cyclodextrin derivative/chromatographic techniques) will be
exploited.

The cyclodextrins are already produced on industrial scale and
they are available in any quantity at drastically reduced prices.
Dimethyl-β-cyclodextrin and the silica bonded cyclodextrins are

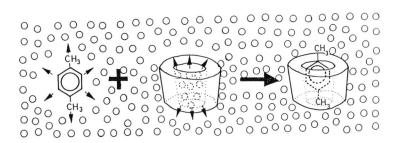

Figure 3. Mechanism of inclusion complexation: ρ-xylene replaces the water molecules of the cyclodextrin cavity.

CHAIN CD POLYMER

NETWORK CD POLYMER

IMMOBILIZED CD

Figure 4. Structural representation of cyclodextrin polymers and immobilized cyclodextrins.

Table I.   Application of Cyclodextrins in Chromatographic Methods

	Thin Layer	Gas-Liquid	Gas-Solid	Gel Inclusion	High Performance Liquid	Affinity	Elektrokinetic
Cyclodextrins (CDs)	M	S			M	M	
Modified CDs		S			M		M
Soluble CD Polymers	M						
Insoluble CD Polymers		S	S				
Immobilized CDs					S	S	

S = in Stationary Phase   M = in Mobile Phase

already available and the industrial production of other derivatives
(polymers) is expected within the next years, after which they too
will be available for analytical purposes.
    In Thin-Layer Chromatography (TLC), the α-cyclodextrin (12-14)
and the soluble β-cyclodextrin polymers have been thoroughly studied
as components of the mobile phase. The γ-cyclodextrin, because of
its excellent solubility, is also promising, especially for larger
molecules, but has not yet been studied. Several attempts have
demonstrated that TLC-plates coated by insoluble finely powdered
cyclodextrin-polymers (i.e. cyclodextrin in stationary phase) can
separate isomers, but this possibility has not yet been fully
exploited.
    In Gas-Liquid Chromatography (GLC), the cyclodextrin (11, 15-18)
(dissolved in appropriate solvent) or cyclodextrin derivatives,
acetylated (19-21) or methylated (10) were found in some cases to
function as highly effective and specific stationary phases. The
β-cyclodextrin polymers were shown to be inadequate (22) or of
limited utility (23) for such purposes.
    In gel inclusion chromatography (GIC), the insoluble, swelling
cyclodextrin polymers are utilized (24-26). For routine analytical
purposes this method is too slow and time consuming, but some highly
effective preparative separations including enantiomeric resolutions
have been published. This approach seems to be very promising for
semi-micro or laboratory scale preparative separations.
    In high performance liquid chromatography (HPLC), the cyclo-
dextrins (12, 27-36) or highly soluble methylated cyclodextrins (37)
in the mobile phase, as well as the silica bonded cyclodextrins (38-
40) as stationary phase have attained spectacular success. A series
of rapid, elegant separations have been published. The field of
application of this method seems to be inexhaustible.

In affinity chromatography (AFC), (41-46) the cyclodextrins are either immobilized (e.g. on Sepharose gel) or are dissolved in the eluent. This method is mainly applied for the purification of amylolytic enzymes.

In electrokinetic chromatography (EKC, electroosmosis, electro-phoresis) the highly soluble ionic cyclodextrin-derivatives (e.g. carboxymethyl-β-cyclodextrin) are employed (47). A non-charged guest will migrate in the electric field, because it is transported by the charged cyclodextrin host.

Next, this paper will review the chromatographic application of soluble cyclodextrin-polymers in thin layer chromatography, and of the insoluble, but swelling β-cyclodextrin bead polymer, in gel inclusion chromatography.

Soluble cyclodextrin polymers in thin-layer chromatography.

Aqueous α-cyclodextrin solutions seem to be generally applicable for TLC separation of a wide variety of substituted aromatics on poly-amide thin-layer stationary sheet (13-14). In most cases, the com-pounds moved as distinct spots and their R_f values were dependent on the concentration of the cyclodextrin in the mobile phase. In a given family of compounds, (o-, m-, and p-nitrophenols, for example) the isomer with the largest stability constant for α-cyclodextrin complex formation had the larger R_f value. In general, the para-substituted isomers have larger R_f values than the meta-isomers, which in turn have larger R_f values than the ortho substituted ones.

An obvious limitation of the application of α-cyclodextrin for a wider variety of compounds is its narrow cavity diameter. Larger molecules do not fit the cavity. Due to its low aqueous solubility, β-cyclodextrin is not adequate for similar purposes. However its highly soluble polymer (a low molecular crosslinked product) proved to be very useful for the TCL separation of larger molecules. The wider cavity diameter, and probably some cooperativeness between the vicinally fixed cyclodextrin-moieties, render such soluble polymers adequate in the mobile phase for a great variety of compounds. The reversed phase TLC-behaviour of antibiotics polymixine (48), 17 substituted s-triazine derivatives (49), 25 triphenyl-methane deriv-atives and analogues (50) 33 nitrostyrene derivatives (51) and 21 barbiturates (52) were studied on silica or cellulose plates.

The utility of the highly soluble β-cyclodextrin derivatives (soluble polymer and dimethyl-β-cyclodextrin) in RPTLC is illustrated in the separation of barbiturates. The lipophilicity of a barbit-urate or any guest decreases when included in a cyclodextrin-cavity. Therefore its mobility is modified in reversed phase thin layer chromatography. With this simple and rapid method, the stability of a complex can be estimated empirically (Table II). The "b" value of the following equation is characteristic for the complex stability (in water:ethanol = 4:1 solution, R_M determined at 5 different cyclodextrin concentrations for 21 barbiturates):

$$R_M = R_{MO} + b.c$$

where R_M = actual R_M values of a compound determined at c(mM) cyclo-dextrin concentrations, $R_{MO} = R_M$ values of a compound extrapolated to zero cyclodextrin concentration, b= decrease of R_M value caused by 1 mM increase of cyclodextrin concentration in the eluent, c= mM cyclodextrin in eluent.

Table II. Structures and the complex stability characterizing "b" values of some barbiturates in 4:1 (v/v) water:ethanol with different cyclodextrins

Barbiturate:

	R_1	R_2	R_3	X	α-CD	β-CD	γ-CD	SCDP	DIMEB
					\multicolumn{5}{c}{"b" values for:}				
3	-$(CH_2)_2CH_3$	-CH_2CH_3	H	O	0.93	2.09	1.33	3.28	
6	-$CH(CH_3)(CH_2)_2CH_3$	-CH_2CH_3	H	O	0.57	1.99	1.86	3.51	
7	-$(CH_2)_2CH(CH_3)_2$	-CH_2CH_3	H	O		2.66	1.39	3.52	
8	-phenyl	-CH_2CH_3	H	O		1.80	1.45	4.90	
10	-cyclohex-3-enyl	-CH_3	CH_3	O		1.56	1.51	2.65	
11	-CH_2CH_3	-CH_2CH_3	H	S			0.94	1.93	
16	-$CH(CH_3)(CH_2)_2CH_3$	-CH_2CH_3	H	S	0.44	2.44	2.85	5.39	9.7
18	-$CH(CH_3)(CH_2)_2CH_3$	-CH_3	H	O	1.38	3.47	2.92	3.82	15.54
19	-$CH(CH_3)CH_2CH_3$	-CH_2CH_3	H	O	0.33	1.03	1.33	2.69	

No.	R₁	R₂							
20	-CH(CH₃)CH₂CH₃	-(CH₂)₃CH₃	H	O	0.36	1.74	1.91	3.65	12.76
21	-CH(CH₃)(CH₂)₂CH₃	-(CH₂)₃-CH₃	H	O	0.62	2.09	1.96	3.94	12.78
22	-(CH₂)₂CH(CH₃)₂	-(CH₂)₃CH₃	H	O	0.39	6.29	1.83	4.94	13.61
23	-(CH₂)₂C((CH₂CH₃)₂(CH₂)₄CH₃	-CH₂CH₃	H	O		8.57	4.37	8.26	30.61
24	-(CH₂)₇CH₃	-CH₂CH₃	H	O	1.8	3.97	2.19	6.52	25.57
25	-CH(CH₃)₂	-CH₂CH=CH₂	H	O	0.22	1.22	1.38	2.45	
26	-CH(CH₃)CH₂CH₃	-CH₂CH=CH₂	H	O		1.02	1.23	2.28	
27	-CH(CH₃)(CH₂)₂CH₃	-CH₂CH=CH₂	H	O	0.5	2.36	2.18	3.85	
28	-2-cyclopentenyl	-CH₂CH=CH₂	H	O	0.39	1.72	1.82	4.15	
29	-1-cyclohexenyl	-CH₂CH₃	H	O	0.40	1.94	1.71	3.91	
30	-CH(CH₃)(CH₂)₂CH₃	-CH₂CH=CH₂	H	S	0.50	3.15	3.4	5.39	24.19
31	-CH(CH₃)CH₂CH(CH₃)₂	-CH₂CH₃	H	O	0.35	3.67	3.14	5.45	15.22
x̄					0.612	2.739	2.033	4.118	17.78
SE (±)					0.437	1.842	0.853	1.521	7.17

According to the "b" values, the α-cyclodextrin form only weak complexes with barbiturates. The stability of β-cyclodextrin and γ-cyclodextrin complexes is similar and higher. Even higher is the stability with the soluble β-cyclodextrin-polymer, the highest being with DIMEB (see the X values in Table II).

From the data the following conclusion can be drawn: Better fitting and higher lipophilicity result in a higher complex stability. Therefore: (i) A longer aliphatic R_1 substituent results in higher stability both with α-cyclodextrin and β-cyclodextrin. (ii) Branching or cyclic R_1 substituent decreases the stability with α-cyclodextrin but increases with the tighter fitting β-cyclodextrin. Unambiguously, an R_1 substituent of unusual size, e.g. an anthracene structure, would not match the β-cyclodextrin cavity, i.e. the increase of R_1 size increases the complex stability only up to an optimum. (iii) The more hydrophobic thiobarbiturates form more stable complexes with β-cyclodextrin which suggests that, at least partially, the heterocyclic ring is also included. The α-cyclodextrin cavity is too narrow for the heterocyclic ring, therefore the X substituent (X = 0, or S) has no significant influence on the α-cyclodextrin complex stability. (iv) The cyclohexenyl ring is more effective than benezene ring.

The results suggest that in α-cyclodextrin-barbiturate complexes the cyclodextrin-cavity includes only R_1 while in β-cyclodextrin complexes both R_1 and part of the pyrimidine moiety are included. This hypothesis does not preclude other interactions between the barbiturate and cyclodextrin molecules. Similar studies were performed on s-triazines, triphenylmethanes and nitrostyrenes.

Separation by inclusion chromatography on cyclodextrin polymers. In 1965 the results were published for the first preparation of insoluble cyclodextrin polymers and their selectivity and superiority in binding various substances (as compared to dextran polymers) (24, 53). Substances which cannot be separated by Sephadex (e.g. o- and m-dichlorobenzene) can be readily separated by β-cyclodextrin polymer. In the presence of cyclodextrin polymers at identical free-substance concentration, the amount of bound substance is much higher (often by 2 orders of magnitude) (see Figure 5). The extent of inclusion follows the Freundlich or Langmuir isotherms (54, 55). For compounds possessing ionizable groups the undissociated forms are predominantly bound.

The guest molecule absorbing capacity of cyclodextrin polymers is significantly higher than would be expected, considering the number of cyclodextrin cavities (56). This was explained by the hypothesis that at least a part of the adjacent cyclodextrin rings bind one guest molecule each, and that between two guest molecules, a third is affixed by hydrophobic forces. The so-called "monofunctional" guest (which interacts with a single cyclodextrin) forms a weaker complex with a substituted or crosslinked cyclodextrin than with a non-derivatized one, probably because of steric hindrance. With "bifunctional" guest, i.e. that interacts with two cyclodextrins, the binding is much stronger (57). This is clearly seen in the values of association constants and on the solubility enhancing effect of β-cyclodextrin and its soluble polymers (Table III).

Table III. Association constant and solubility enhancement of
"monofunctional" and "bifunctional" guests with CD
and soluble CD polymer

Guest	$K_a (x\ 10^{-3})$		$S/S_o{}^a$
	β-CD[b]	S-β-CDP[b]	
monofunctional: m-chlorobenzoic acid	2.2	0.5	2.1
bifunctional: 4-dimethylaminoazo-benzene	0.35	7.0	40.0

$^a S_o$ refers to the solubility in water.

$^b \beta$-CD = 4.0×10^{-3} M,   S-β-CDP = 5.6×10^{-3} M (as CD unit).

A serious limitation of the chromatography on cyclodextrin polymers is that it can be performed only in aqueous solutions. The
retention of the guests depends on the stability of the complex and
changes with the polarity, hydrophobicity, size and geometry of the
guest molecule and the size of the internal cavity of the cyclodextrin molecules. Moreover, it changes further with the temperature
and other experimental conditions (e.g. pH and the composition of
the mobile phase). Secondary effects, such as gel permeation and
weak adsorption, can also interfere with the complexation. In
favorable cases, these effects jointly increase the chromatographic
separation.

Figure 6 shows the complete separation of five amino acids
(lysine, alanine, phenylalanine, tyrosine and tryptophan) on a
column packed with cyclodextrin polymers (58). The best separation
of these amino acids was obtained on the column packed with β-cyclo-
dextrin polymer. On the other hand, tryptophan could be separated
on α-cyclodextrin polymer column with the best selectivity. Also,
fifteen additional non-aromatic natural α-amino acids were chromat-
ographed on β-cyclodextrin polymer, but their peaks appeared either
between, or together with, those of alanine and lysine.

The study of the chromatographic behavior of natural indole
alkaloids on cyclodextrin polymers was different, and unexpectedly
high retentions were observed in mildly acidic buffer solutions at
room temperature, which permitted their separation by inclusion
chromatography (25) (Table IV). Figure 7 shows the separation of
two Vinca-alkaloids of very similar structure, the (+)-vincamine
and (+)-apovincamine.

In favorable cases, not only structurally related compounds or
isomers can be separated, but also enantiomers, when the stability of
the diastereomer complexes (CD.G) of the guest (G) is different, i.e.
$K_{(+)} \neq K_{(-)}$:

$$(+)CD + (+)G \rightleftharpoons (+)CD.(+)G$$
$$(+)CD + (-)G \rightleftharpoons (+)CD.(-)G$$

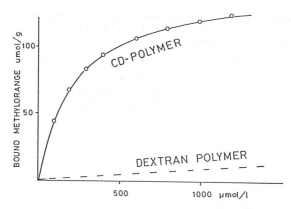

Figure 5. Comparison of the degree of binding of methylorange to
dextran polymer and by β-cyclodextrin polymer in aqueous solution.

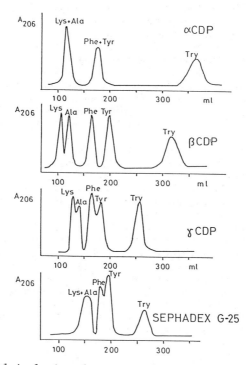

Figure 6. Gel inclusion chromatographic separation of amino acids
on α-, β-, γ-cyclodextrin bead polymer columns, and on Sephadex
G-25 column (1.6x88 cm, pH 5.0 phosphate buffer, flow rate 10
ml/h, 20 °C).

Table IV.   Separation of some indole alkaloids on β-CD polymer
column with phosphate buffer at pH 5.5 as mobile phase

Alkaloid	V_e/V_t enantiomer (-)	(+)	Selectivity factor
Vincadifformine	6.0	4.2	1.43
Aspidospermidine	1.58	1.73	1.09
Quebrachamine	4.75	3.98	1.19
N-Methylquebrachamine	3.48	3.63	1.04
Vincadine	1.82	1.89	1.04
Apovincamine	3.85	3.58	1.08
Eburnamonine	5.6	5.3	1.06
Vincamine	1.69	1.72	1.02

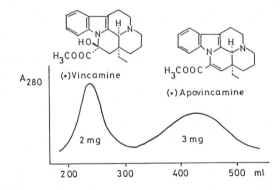

Figure 7.   Separation of (+)-vincamine (2 mg) and (+)-apovincamine
(3 mg) on β-cyclodextrin polymer (1.6x90 cm, pH 5 citrate buffer,
flow rate 80 ml/h, 20 °C).

The chances of the chromatographic resolution by inclusion chromatography were systematically studied on a series of enantiomer pairs
of indole alkaloids as model compounds (e.g. Table V) and promising
results were achieved on both analytical and preparative scale (26).

Table V. Specific elution volumes (V_e/V_t) and selectivity factors of
          vincadifformine enantiomers on βCD polymer column

Particle size (μm)	Eluent buffer		V_e/V_t enantiomer (-)	V_e/V_t enantiomer (+)	Selectivity factor
90-125	Citrate,	pH 4.0	2.8	2.5	1.12
	Phosphate,	pH 5.0	3.1	2.6	1.19
	Phosphate,	pH 5.0	4.6	3.3	1.39
63-90	Phosphate,	pH 5.0	3.3	2.5	1.32
	Phosphate,	pH 5.5	6.0	4.2	1.43
	Citrate,	pH 5.5	4.7	3.5	1.34

Figure 8 shows the analytical base-line separation of quebrachamine
antipodes by inclusion chromatography on β-cyclodextrin polymers.
    Until now, almost exclusively analytical works have been pub-
lished, i.e. several mg racemic mixtures were separated without
isolation of the enantiomers. Because cyclodextrin-polymers are
not yet industrially produced, their accessibility is limited,
particularly in quantities which are needed for preparative columns.
This field however seems to be promising, because the production and
availability in satisfactory quantities of β-cyclodextrin bead polymers is expected within the next few years.
    The preparative chromatography of 500 mg racemic (+)-vincadiff-
ormine on β-cyclodextrin polymer can be seen in Figure 9. From the
chromatographic fraction, 230 mg crude (+)-vincadifformine was
isolated. Its optical purity after recrystallization was 98.3 %. The
other fraction was 245 mg crude (-)-vincadifformine. After recrystallization its optical purity was 92.5 %. These are excellent results,
particularly considering the 92 % yield of the crude product. The
loading capacity of the preparative column was tested by increasing
the amount of the racemic mixture. Separation was achieved at higher
loadings, but as expected the optical purity of both of the enantiomer products somewhat decreased. For example, resolving 800 mg
racemic (+)-vincadifformine under the same circumstances as above
gave 350 mg (87.5 %) crude (+)-vincadifformine from the first eluate
fraction, and 380 mg (95 %) crude (-) enantiomers. The optical
purities (after recrystallization) were 81.6 % and 87.5 % respectively.

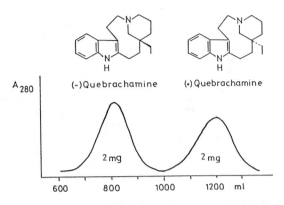

Figure 8. Baseline resolution of (+)-quebrachamine (2 mg) and (-)-quebrachamine (2 mg) on β-cyclodextrin polymer (1.6x85 cm, pH 6.8 phosphate buffer, flow rate 50 ml/h, 20 °C).

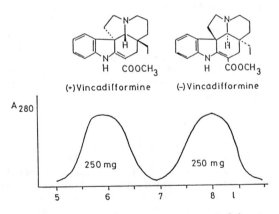

Figure 9. Resolution of a racemic mixture of (+)-vincadifformine (250 mg) on a preparative β-cyclodextrin polymer column (5x90 cm, pH 5.5 phosphate buffer, flow rate 300 ml/H, 20 °C).

The ability of some components of nucleic acids, especially those with an adenine base, to form complex with β-cyclodextrin, can also be readily used for chromatographic separations of various nucleotides and nucleosides (59). A substantial problem associated with application of cyclodextrin polymer gels, is that the accessibility of the cyclodextrin cavities on the surface and within the interior of the polymer particle is rather different. The rate of entrapment and release of solutes from the streaming liquid is obviously a diffusion controlled process. Consequently, a longer time is needed to reach an equilibrium within the particle than on its surface. The accessibility of the cyclodextrin rings will be more uniform, if the cyclodextrin is immobilized on the surface of non-complexing polymer particles (polyacrylamide, agarose (60,61) cellulose (62), and silica (63)). Therefore, a better separation (however lower capacity) is expected.

Columns in which β-cyclodextrin was immobilized on polyacrylamide or agarose gel were shown to be very useful in the separation of disubstituted benzene isomers (60,61). Acetylating the immobilized β-cyclodextrin further improves the selectivity, i.e. it can completely separate o-, m- and p-toluidine, and dinitrobenzenes (64) which cannot be done on unmodified stationary phase.

Not only analytical or preparative separations can be performed on cyclodextrin polymer columns, but also undesired components can be removed from aqueous solutions, bitter tasting substances (naringin, limonin) can be removed or at least their concentration can be strongly reduced after treatment of citrus juice with cyclodextrin polymers in batch or column process (65,66). Phenylalanine can be eliminated from dietetic protein hydrolysates (67), water-soluble organic substances (e.g. polychlorinated biphenyls (68), 2-naphtalenecarboxylate or phenol can be removed from aqueous solutions (e.g. from pharmaceutical wastewater) by polystyrene-cyclodextrin derivatives (69), by β-cyclodextrin immobilized on cellulose (70) or by β-cyclodextrin-polyurethane polymer (71).

Concluding remarks

The selectivity of cyclodextrins toward the various molecules is not high enough to attain complete (or acceptable) separations by one-step operations. Enrichment, of one component or partial separation of various components of a mixture can be attained relatively easily. However using cyclodextrins in multistep processes, i.e. the various chromatographic techniques, very effective separations can be achieved. Particularly in RP-HPLC the application of immobilized CDs and CDs dissolved in the mobile phase became one of the most promising methods.

In the coming years the modified cyclodextrins may bring about even more specific separations, or separation of poorly soluble drugs in aqueous systems. The most challenging aim is the preparative separation of mixtures, and resolution of racemates by the immobilized chiral CDs, and scaling-up such methods to industrial technologies. The CDs are already accessible at reasonable price, the same is expected for modified CDs and CD polymers in the coming years. It is hoped, that such CD-derivatives will achieve similar significance in the separation technology, than in the analytical chemistry.

Literature Cited

1. French D., <u>Adv. Carbohydrate Chem.</u> 1957, 12,189.
2. Bender M. I., Komiyama M. "Cyclodextrin Chemistry", Springer Verlage, Berlin-Heidelberg-New York, 1978.
3. Szejtli J. "Cyclodextrin and their Inclusion Complexes" Akademiai Kiado, Budapest, 1982.
4. Szejtli J. in "Inclusion Compounds" (Ed. Atwood J. L. Davies J.E.D., MacNicol D.) Academic Press, London 1984 Vol.3.p.33.
5. Croft A., Bartsch R. A., <u>Tetrahedron</u> 1983, 39, 1417, Chem. Abstr. 99, 22770.
6. Szejtli J., Fenyvesi E., Zoltan S., Zsadon B., Tudos F. US Patent 4,274,985, 1980.
7. Hirai H., Komiyama M., Yamamoto H., <u>J. Inclusion Phenomena</u> 1985, 2 655.
8. Lee C.H., <u>Sep. Sci. Technol.</u>, 1981, 16, 25, <u>Chem. Abstr.</u> 94, 36979.
9. Lee C.H., <u>J. Appl. Polym. Sci.</u> 26, 96, <u>Chem. Abstr.</u> 123470.
10. Smolkova-Keulemansova E., <u>J. Chromatogr.</u> 1982, 251, 17.
12. Hinze W. L., <u>Separations and Purification Methods</u>, 1981, 10, 159.
13. Burkert W. G., Owensby C. N., Hinze W. L., <u>J. Liq. Chromatogr.</u>, 1981, 4, 1065.
14. Hinze, W. L., Armstrong D. W., <u>Anal. Lett.</u> 1980, 13, 1093.
15. Smolkova-Keulemansova E., Feltl L., Kiysi J., <u>J. Inclusion Phenomena</u> 1985, 3, 183.
16. Koscielski T., Sybilska D., Feltl L., Smolkova-Keulemansova E., <u>J. Chromatogr.</u> 1984, 286, 23.
17. Koscieiski T., Sybilska D., Belniak S., Jurczak J., <u>Chromatographia</u> 1984, 19, 292.
18. Koscieiski T., Sybilska D., Jurczak J., <u>J. Chromatogr.</u>, 1983, 280, 131.
19. Sand D. M., Schlenk H., <u>Anal. Chem.</u> 1961, 33, 1624.
20. Schlenk H., Gellerman J. L., Sand D. M., <u>Anal. Chem.</u> 1962, 34, 1529.
21. Tanaka M., Kavano S., Shono T., <u>Fresenius Z. Anal. Chem.</u> 1983, 316, 54, <u>Chem. Abstr.</u> 99, 205314.
22. Cserhati t., Dobrovolszky A., Fenyvesi E., Szejtli J., <u>J. High REsolut. Chromatogr.</u> 1983, 6, 442, <u>Chem. Abstr.</u> 99, 201022.
23. Scypinski S., Love Cline L. J., <u>Anal. Chem.</u> 1984, 56, 331.
24. Solms J., Egli R. H., <u>Helv. Chim. Acta</u> 1965, 48, 1225.
25. Zsadon B., Szilasi M., Tudos F., Szejtli J., <u>J. Chromatogr.</u> 1980, 208, 109.
26. Zsadon B., Decsei M., Szilasi M., Tudos F., Szejtli J., <u>J. Chromatogr.</u> 1983, 270, 127.
27. Uekama K., Hirayama F., Ikeda K., Inaba L., <u>J. Pharm. Sci.</u>, 1977, 66, 706.
28. Gazdag M., Szepesi G., Huszar L., <u>J. Chromatogr.</u> 1986, 351, 128.
29. Debowski J., Sybilska D., Jurczak J., <u>Chromatographia</u> 1983, 282, 83.
30. Debowski J., Sybilska D., Jurczak J., <u>J. Chromatogr.</u> 1982, 237, 303.
31. Debowski J., Sybilska D., Jurczak J., <u>Chromatographia</u> 1982, 16, 198.

32. Debowski J., Jurczak J., Sybilska D., Zukowski J., J. Chromatogr.
    1985, 206, 329.
33. Sybilska D., Debowski J., Jurczak J., Zukowski J., J. Chromatogr.
    1984, 286, 163.
34. Sybilska D., Lipkowski J., Woycikowski J., J. Chromatogr. 1982,
    253, 95.
35. Nobuhara Y., Hirano S., Nakanishi Y., J. Chromatogr., 1983, 258,
    276.
36. Sybilska D., Zukowsky J., J. Liq. Chrom. in press, 1986.
37. Tanaka M., Miki T., Shono T., J. Chromatogr. 1986, 330, 253,
    Chem. Abstr. 194, 45127.
38. Armstrong D. W. US. Patent 4,539,393, 1985.
39. Beesley T. E., Am. Lab. 1985, 17,78,80, 83-7, Chem. Abstr. 103,
    31790.
40. Tanaka M., Kawaguchi Y., Niinae T., Shono T., J. Chromatogr.
    1984, 314, 193, Chem. Abstr. 102, 75056.
41. Vretblad P., FEBS Lett., 1975, 47, 86, Chem. Abstr. 82, 53181.
42. Silvanovich M.P., Hill R. D., Anal. Biochem. 1976, 73, 430.
43. Weselake R. J., Hill R. D., Carbohydr. Res. 1982, 108, 153.
44. Hoschke A., Laszlo E., Hollo J., Starch 1976, 28, 426.
45. Ludwig I., Ziegler I., Beck E., Plant Physiol. 1984, 74, 856.
46. Laszlo E., Banky B., Seres G., Szejtli J., Starch 1981, 33, 281.
47. Terabe S., Ozaki H., Otsuka K., Ando T., J. Chromatogr. 1986,
    332, 211, Chem. Abstr. 104, 45135.
48. Cserhati T., Bordas B., Fenyvesi E., Szejtli J., J. Chromatogr.
    1983, 259, 107.
49. Cserhati T., Bordas B., Fenyvesi E., Szejtli J., J. Inclusion
    Phenomena 1983, 1, 53.
50. Cserhati T., Oros Gy., Fenyvesi E., Szejtli J., J. Inclusion
    Phenomena 1984, 1, 395.
51. Cserhati T., Bordas B., Kis-Tamas A., Mikite Gy., Szejtli J.,
    Fenyvesi E., J. Inclusion Phenomena 1986, 4, 55.
52. Szejtli J., Cserhati T., Bordas B., Bojarski J., To be
    published, 1986.
53. Szejtli J., Fenyvesi E., Zsadon B., Starch 1978, 30, 127.
54. Wiedenhof N., Trieling R. G., Starch 1971, 23, 129.
55. Wiedenhof N., Starch, 1969, 21, 163.
56. Lammers J. N., Van Diemen A. J., Rec. Trav. Chim. Pays-Bas
    1972, 91, 733, Chem. Abstr. 77, 7476.
57. Harada A., Furue M., Nozakura S., Polym. J. (Tokyo); 1981, 13,
    777, Chem. Abstr. 96, 218351.
58. Zsadon B., Szilasi M., Fenyvesi E., H. Otta K., Szejtli J.,
    Tudos F., Acta Chim. Hung. 1979, 100, 265.
59. Hoffman J. L., J. Macromol. Sci. Chem. 1973, A7, 1147.
60. Tanaka M., Mizobuchi Y., Sonoda T., Shono T., Anal. Lett. 1981,
    14, 281, Chem. Abstr. 94, 198116.
61. Tanaka M., Kawaguchi Y., Shono T., Nakae M., Mizobuchi Y.,
    J. Chromatogr. 1982, 246, 207.
62. Szejtli J., Zsadon B., Fenyvesi E., Horvath K., Tudos F.,
    Brit. Pat., 1982, GB 2,083,821.
63. Tanaka M., Kawaguchi Y., Nakae M., Mizobuchi Y., Shono T.,
    J. Chromatogr. 1984, 299, 341, Chem. Abstr. 101, 210493.
64. Tanaka M., Kawaguchi Y., Shono T., J. Chromatogr. 1983, 267,
    285.

65. Shaw P. E., Wilson G. W., J. Food Sci., 1983/9, 48, 6466,
    Chem. Abstr. 98, 17776.
66. Shaw P. E., Tatum J. H., Wilson, C. W., J. Agric Food Chem.
    1984, 32, 832, Chem. Abstr. 101, 53605.
67. Specht M., Rother M., Szente L., Szejtli J., Ger. Patent (DDR),
    1980, 147, 615.
68. Friedman R. B., Gottneid D. J., Mauro D. J., Owen R. L.,
    West I. R., 190th ACS National Meeting, Chicago, 1985.
69. Tabuse I., Shimizu N., Yamamura K. Jpn. Kokai 1979, 79,60,76,
    Chem. Abstr. 91, 162667.
70. Otta K., Fenyvesi E., Zsadon B., Szejtli J., Tudos F.,
    Proc. I. Inst. Symp. Cyclodextrins, (Ed.: J. Szejtli), Reidel
    Dordrecht, 1982. p.357.
71. Kawaguchi, Y.; Mizobuchi, Y., Tanaka, M., Shono, T., Bull. Chem.
    Soc. Jpn. 1982, 55, 2611, Chem. Abstr. 97, 101, 329, Eur. Pat.
    Appl. 60532, Chem. Abstr. 98, 91993.

RECEIVED May 11, 1987

Chapter 12

Cyclodextrins as Mobile-Phase Components for Separation of Isomers by Reversed-Phase High-Performance Liquid Chromatography

Danuta Sybilska

Institute of Physical Chemistry, Polish Academy of Sciences, Kasprzaka 44/52, Warsaw 01-224, Poland

The use of cyclodextrins as the mobile phase components which impart stereoselectivi-
ty to reversed phase high performance liquid chromatography (RP-HPLC) systems are
surveyed. The exemplary separations of
structural and geometrical isomers are presented as well as the resolution of some
enantiomeric compounds. A simplified scheme
of the separation process occurring in RP-
HPLC system modified by cyclodextrin is
discussed and equations which relate the capacity factors of solutes to cyclodextrin
concentration are given. The results are
considered in the light of two phenomena
influencing separation processes: adsorption
of inclusion complexes on stationary phase and complexation of solutes in the
bulk mobile phase solution.

The most characteristic property of cyclodextrins(CD's)
is their remarkable ability to form molecular inclusion
compounds with various organic and inorganic species of
neutral or ionic nature (1). Their growing significance
in chromatography (2,3) arises from the fact, that CD
complexation meets almost all the main requirements of
this method set up on the adapted process.
    First, CD complexation is selective, moreover high-
ly stereoselective. CD inclusion processes are influenced mainly by hydrophobicity and shape of guest (G) molecule i.e. by the fit of the entire or at least part
of complexed molecule to the CD (host) cavity.Thus steric factors are of crucial importance for CD inclusion
compounds formation and their stability. For that
reason CD inclusion can be considered as a procedure

0097-6156/87/0342-0218$06.00/0

of choice for separation of isomers i.e. in the field
where many analytical and preparative problems remain
unsolved or only partly solved.

Second, CD complexation processes occurring in solution are reversible; hysteresis phenomena are not
observed. The process of equilibration in the solution
is relatively fast, the rate constants of complexation
are usually of the same order as those of diffusion
controlled processes.

Third, CD's are stable within a large range of pH,
they are resistive to the light and they do not absorb
in the full UV range commonly used in chromatographic
detection. Moreover CD's are not toxic.

First of all CD complexation has been used to advantage in classical liquid chromatography and thin
layer chromatography.These studies procured very interesting and valuable results which have been recently
reviewed (2,3). However, the columns usually containing
polymers with incorporated CD molecules are of very low
efficiency, owing to the complex mechanism of sorption
involving both gel permeation and molecular inclusion.

For the application of stereoselective processes
of CD inclusion in high performance liquid chromatography (HPLC) two different approaches have been recently
designed: the use of chemically bonded CD silica stationary phases (4-16) and the application of CD's as
the mobile phase components in reversed phase (RP)
systems (17-28).

The results obtained by the first method especially
those of recent studies performed with a commercially
available stable α-, β- and γ-CD bonded phases (8 -
15) and with their newest improved modifications (16)
demonstrate the great practical value of the sorbents
and procedure.These studies dealt with structural and
geometrical isomers and diastereoisomers as well as
enantiomers of numerous compounds of various hydrophobic or hydrophilic nature.

In this chapter, attention will be focused on the
studies that utilize CD's as the mobile phase modifiers.
It should be noted, that two significant facts found
earlier, opened the route to the idea of using CD's as
the mobile phase components in RP-HPLC systems.Namely:
1) CD's dissolved in the mobile phase solutions have
been used in thin-layer chromatography with polyamide
stationary phase (29,30) and 2) CD complexation equilibria of ionic compounds were studied by a chromatographic method using an ion exchanger as stationary phase
and mobile phase solutions containing CD's in various
concentrations (31).

Equilibria and Equations.

RP systems containing CD's in mobile phase solutions may
involve many species of the solute: neutral, ionic, free
or bound to one or more CD molecules. Consequently the
adsorption and complexation equilibria are complicated.
Supposing that only one species of the solute, i.e.
neutral molecules (G) , is present in the solution and
takes part in the processes of adsorption and complexation, complexes of 1:1 stoichiometry are exclusively
formed and CD does not influence the properties of the
RP stationary phase, we then obtain the following
simplified scheme for description of the equilibria:

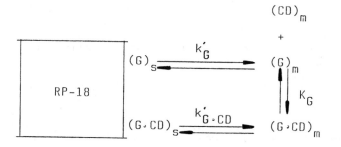

where G stands for guest molecules (solute), the subscripts s and m denote the stationary and mobile phase
respectively; K_G is stability constant of the (G·CD)
complex and k'_G and $k'_{G·CD}$ are capacity factors of the
free G molecule and its G·CD complex, respectively. For
such a system the apparent capacity factor (k') can be
expressed as follows:

$$k' = \frac{k'_G + k'_{G·CD} K_G [CD]_m}{1 + K_G [CD]_m} \qquad (1)$$

Equation 1 describing a simple RP system (18,23) is analogous to that derived for the first time by Uekama
et al (31) for determination of the stability constants
of CD complexes with various ionic species by ion exchange chromatography. The analogous equations have been
proposed by Horvath et al (32) for ion pair chromatography.
   Equation 2 (23) arises from Eqation 1 by simple
transformation for linearization:

$$k' = \frac{k'_G - k'}{K_G [CD]_m} + k'_{G·CD} \qquad (2)$$

The aqueous mobile phase solutions must frequently contain not only CD but an additional organic solvent whose molecules are also included in the CD cavities(33). The competitive influence of organic solvent on complexation equilibria may be expressed by Equation 3 (23) for the apparent CD molar concentration $[CD]_m$ being smaller than the overall molar concentration $[CD]_m^o$:

$$[CD]_m = \frac{[CD]_m^o}{K_{solv}[solv]_m^o + 1} \tag{3}$$

where $[solv]_m^o$ is the initial molar concentration of organic solvent and K_{solv} is the stability constant of the 1:1 CD inclusion complex with an organic solvent molecule. Equation 3 was derived on the assumption that $[solv]_m^o$ is very similar to the equilibrium concentration of organic solvent, what seems to be reasonable under the condition:

$$[solv]_m^o \gg [CD]_m^o$$

For weak acids and bases, i.e.,substances undergoing dissociation, Equation 4 (more complicated than Equation 1) was derived (18) to evaluate how pH and CD concentration affect their retention on RP columns. It takes into account the acid-base and complexation equilibria of both neutral and ionic species as well as the adsorption of all species on the stationary phase.

$$k' = \frac{k_G' + k_{iG}'K_a/[H^+] + k_{G \cdot CD}'K_G[CD] + k_{iG \cdot CD}'K_{iG}[CD]K_a/[H^+]}{1 + K_a/[H^+] + K_G[CD] + K_{iG}[CD]K_a/[H^+]} \tag{4}$$

where K_a is the acidity constant,K_G and K_{iG} are stability constants of CD complexes of neutral and ionic species respectively and $k_G', k_{iG}', k_{G \cdot CD}'$ and $k_{iG \cdot CD}'$ are capacity factors of unionized,ionized,complexed-unionized and complexed-ionized forms of solute respectively.

Experimental Verification. The changes of capacity factor values k' with β-CD concentration in the mobile phase solution are illustrated in Figure 1 as the behavior of methylphenobarbital enantiomers on RP-18 columns (26). The similar influence of CD on k' values was observed for all the studied compounds (disubstituted benzenes, mandelic acid and its derivatives, some aromatic aminoacids, some barbiturates and hydantoins)(17-19,21-26,28). α - or β-CD additions were always followed by a decrease in the apparent capacity factor (k') values. These results suggest that the adsorption

of CD complex on RP-18 phase is always smaller than that
of the corresponding free molecule if the assumption that
CD does not change RP stationary phase is certainly valid

$$k_G' > k_{G \cdot CD}'$$

Generally, in the case of β-CD the determined k' values
satisfy the linear relation

$$k' \text{ vs } (k_G'-k')/[\![\beta\text{-CD}]\!] \text{ from Equation 2}$$

which proves that in this case all three main assumptions
of the scheme given above and Equation 1 are approximati-
vely true. This linear relationship enabled evaluation of
the stability constants K_G and the capacity factors
$k_{G \cdot \beta\text{-CD}}'$ for β-CD complexes with various compounds. Some
of these values are collected in Table I (22,23,26,34).
The data for cresols, hexobarbital and mephenytoin were
evaluated by the least-squares method using Equation 2.
The data concerning mandelic acid and para-nitrocinnamic
acid were determined from Equation 4 by expanding the
non-linear function $k' = k([H^+][CD])$ into a Taylor series
and applying a numerical procedure (18)
    A few small deviations observed in the case of β-CD
behaviour (e.g. negative values obtained sometimes for
$k_{G \cdot \beta\text{-CD}}'$) may be attributed to two phenomena: either β-CD
influences, to some extent, hydrophobic properties of RP
stationary phase or two complexation stages occur. The
very poor confirmation of Equation 1 observed for α-CD
points out the bigger divergences between its real behaviour in a RP system and the main assumptions of the
scheme given above. For that reason no further evaluation
of stability and adsorption of α-CD complexes was attemp-
ted.

Optimization. The data discussed above (equilibria,
equations, Table I) lead to the conclusion that the
resolution of the compounds achieved in RP systems modified by CD may be due to two various phenomena. It may
arise from
- differences of stability constants of β-CD complexes
  (K_G),
-and difference in adsorption of β-CD complexes
  ($k_{G \cdot \beta\text{-CD}}'$) on the stationary phase.
These two factors may influence resolution of isomers
in an additive (e.g. mandelic acid enantiomers, ortho
and meta cresols) or substractive manner (e.g. mephenytoin enantiomers). Thus for the designing and optimization of resolution one should know the stability constants and adsorption properties of CD complexes with
the compounds being separated.
    Two examples of the optimization of separation
factor values for enantiomers are presented in Figures 2

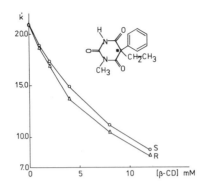

Figure 1. Plots of capacity factors k' vs β-CD concentration for methylphenobarbital enantiomers. Stationary phase: 10 μm LiChrosorb RP-18; mobile phase: 20% ethanol–buffer solution of pH=2.0 containing various β-CD concentrations; temperature: 25 °C. (Reprinted with permission from ref. 26. Copyright 1986 Marcel Dekker.)

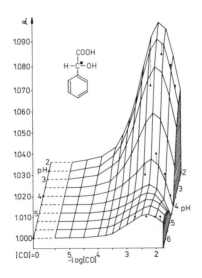

Figure 2. Selectivity factor values calculated for mandelic acid enantiomers $\alpha_{L/D}$ as a function of pH and log [β-CD]; • denotes $\alpha_{L/D}$ value from direct measurements. (Reprinted with permission from ref. 34. Copyright 1984 Polska Akademia Nauk Instytut Chemii Fizycznej.)

Table I. Capacity factors of solutes (k'_G), calculated capacity factors of their β-CD complexes ($k'_{G\cdot\beta-CD}$) and stability constants (K_G) of β-CD complexes of cresoles, p-nitrocinnamic acids and enantiomers of mandelic acid, mephenytoin and hexobarbital. Stationary phase: 10 μm LiChrosorb RP 18

Compounds	Solvent	k'_G	$k'_{G\cdot\beta-CD}$	K_G	k'_{iG}	$k'_{iG\cdot\beta-CD}$	K_{iG}
ortho cresol	20 vol % ethanol in H$_2$O	15	2	34			
meta cresol		15	1	41			
para cresol		15	1	69			
cis	buffer aqueous solutions	308	28	1780	30	14	795
trans		713	0	1750	58	11	984
para nitro cinnamic acid							
mandelic acid	buffer aqueous solutions						
L - (+)			4.4	540		1.40	203
D - (-)			4.3	607		1.37	207
mephenytoin	20 vol % ethanol in aqueous buffer of pH 2.0						
S		15	2.0	43			
R		15	5.9	49			
hexobarbital							
S		18	-0.4	131			
R		18	-0.4	151			

and 3. Equation 4 and the data from Table I for mandelic acid enantiomers were used to examine the relationship between the selectivity factor $\alpha_{L/D}$ and CD concentration as well as pH of the mobile phase solution. The results of this procedure are shown in Figure 2 (34). It is seen that the best separations of mandelic acid enantiomers should be achieved on RP-18 column with more acidic solutions (of pH $\leqslant$ 2) and at the moderate concentrations of β-CD. However such acidic media are not advisable for LiChrosorb RP-18 columns, also the CD's can undergo acidic hydrolysis too at low pH.

Plots of selectivity factor (calculated using Equation 2 and the data from Table I) for mephenytoin and hexobarbital enantiomers versus CD concentration are shown in Figure 3 a,b (35). The profiles of relation α vs [β-CD] for these two compounds are different because two different factors determine resolution of their enantiomers: difference in K_c values for hexobarbital and difference in k_{c}, β-CD values for mephenytoin. The latter case represents an interesting example: the resolution of its enantiomers arises from the great differentiation in the adsorption of diastereoisomeric β-CD complexes. The calculated selectivity factor α for these complexes is ca 3 (see Table I). In this particular case selectivities of the two processes: adsorption and complexation in the bulk mobile phase solution are opposite to each other; enantioselectivity arising from selective adsorption dominating over differentiation in the solution. Unfortunately the stabilities of diastereoisomeric β-CD·mephenytoin complexes are relatively small and solubility of β-CD in the mobile phase solution is rather limited. Therefore one cannot shift the complexation equilibrium

$$\beta\text{-CD} + \text{mephenytoin} \rightleftharpoons \beta\text{-CD}\cdot\text{mephenytoin}$$

towards prevailing formation of the inclusion complex through increasing the β-CD concentration. In consequence of this, the mentioned above optimal separation factor α equal ca 3 is not available experimentally.

Examples of Separations.

Structural Isomers. Chromatograms illustrating the separation of ortho, meta and para isomers of cresol (23) and and xylene (28)on RP columns are shown in Figures 4 and 5. They enable a comparison of the chromatographic properties and selectivities due to α- and β-CD complexation between positional isomers of the above compounds. Similar behaviour was observed for ortho,meta and para isomers of fluoronitrobenzene, chloronitrobenzene, iodonitrobenzene, nitrophenol, nitroaniline, dinitrobenzene (23), nitrocinnamic acid (22) some mandelic acid derivatives (19,21,34) and ethyltoluene (28). Both α-CD and

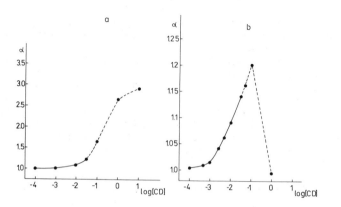

Figure 3. Selectivity factor values calculated for (a) mephenytoin ($\alpha_{R/S}$) and (b) hexobarbital ($\alpha_{S/R}$) enantiomers as a function of log [β-CD]. The full line corresponds to the conditions experimentally available.

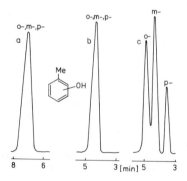

Figure 4. Separation of cresols on 10-μm LiChrosorb RP-18 column (150 x 4.5 mm i.d.) (a) without CD, (b) with 3×10^{-2} M α-CD, and (c) with 2×10^{-2} M β-CD. Solvent composition: 20 vol% ethanol in water; flow rate: 2.4 mL/min; temperature: 20 °C. (Reprinted with permission from ref. 23. Copyright 1985 American Chemical Society.)

β -CD form inclusion complexes with disubstituted ben-
zenes.Nevertheless, in most cases only β -CD complexation permits effective separations of positional isomers
of disubstituted benzenes. The only exception from this
regularity, so far observed, is nitrobenzoic acid; its
ortho, meta and para isomers were more efficiently separated with α -CD solutions (18) than with β -CD.
    The sequence of elution of positional isomers of
halogen derivatives of nitrobenzene (23) ,xylene and
ethyltoluene (28) from RP column modified with a β -CD
solution is identical: 1) ortho, 2) para and 3) meta.
It was mentioned earlier that these selective chromatographic separations, achieved with CD solutions, are
due to the difference in stability constants of inclusion
complexes in the mobile phase solution and to the difference in the adsorption of these complexes on RP phase.
As the capacity factors of β-CD complexes of disubstituted benzene derivatives are comparatively low and their
differentation is also rather small - thus the final
observed selectivity is mainly determined by the differences in the stability constants between β-CD complexes
of ortho, meta and para isomers. The sequence of elution
of ortho, meta and para isomers from RP column should be
therefore reverse to the stabilities of their β-CD complexes. This sequence should be opposite to that observed
on the columns filled with β-CD silica bonded phases
(for the same compounds and under similar conditions).
In fact, the results concerning β-CD activity in mobile
phase solutions, concerning positional isomers of disubstituted benzene derivatives (23,28) seem to be consistent with those obtained by other authors using β-CD
silica bonded phases (4-8). It should be mentioned that
all of the above considerations concern mainly neutral
molecules; for the compounds undergoing dissociation, the
equilibria are more complicated and the sequence of elution of isomers more variable.
    Chromatograms in Figure 6 show the separation of tri-
methylbenzenes. As it was observed for dialkylbenzenes
β -CD complexation not only improves selectivity towards
trimethylbenzene isomers, but also works as an organic
solvent by lowering their capacity factors. This makes
the time of analysis shorter and detectability better
(28). The improvement in the resolution of trimethyl
benzenes due to the α -CD complexation is not so obvious.

Geometrical Isomers. Figure 7 shows the chromatogram of
a mixture of all six isomers of nitrocinnamic acid as an
example of the separation of geometrical isomers. In this
case, however, the main difficulty lies not in the sepa-
paration of cis from trans isomers, but in the resolution
of meta from para substituted compounds whether in cis or
trans configuration.

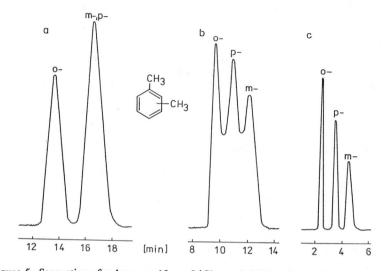

Figure 5. Separation of xylenes on 10-μm LiChrosorb RP-8 column (50 × 4.0 mm i.d.)
(a) without CD, (b) with 3.10^{-2} M α-CD, and (c) with 2.7 × 10^{-2} M β-CD. Solvent
composition: 20 vol% ethanol in water; flow rate: 3.6 mL/min; temperature: 25 °C.
(Reprinted with permission from ref. 28. Copyright 1986 Elsevier Science Publishers.)

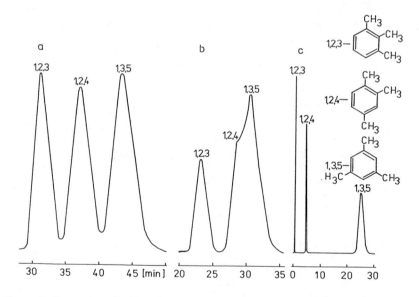

Figure 6. Separation of trimethylbenzenes (a) without CD, (b) with 3.10^{-2} M α-CD,
and (c) with 2.7 × 10^{-2} M β-CD. Conditions as in Figure 5. (Reprinted with
permission from ref. 28. Copyright 1986 Elsevier Science Publishers.)

Chiral Compounds. As CD's are composed of D-glucose
they are chiral. CD complexation represents therefore
a potential tool for separation of other chiral compounds into enantiomers. In fact various interesting
separations of chiral compounds into enantiomers have
been achieved using β-CD silica stationary phases (8,9,
12-16).
        In this matter the surveyed method has been exempli-
fied at first by the resolution of mandelic acid enantiomers (17). The further studies concerned the chiral recognition of mandelic acid derivatives substituted in the
side chain and/or in the aromatic ring (19,21). It has
been found that the enantioselectivity arising from inclusion in β-CD molecules is distinguishable only for compounds containing at the chiral carbon atom an intact
carboxylic group and another polar group (e.g. OH, NH$_2$)
able to form hydrogen bond. It was additionally assumed
that the insertion of the phenyl group in the central
cavity of β-CD provides the third point of contact,
indispensable for achieving enantioselectivity in a chromatographic system, according to the "three points of
attachment" concept of Dalgliesh (36). This assumption
seems to be confirmed by chloromandelic acids behavior.

Table II. Capacity Factors k´ of Enantiomers of
Mandelic Acid and its Chloroderivatives Determined
on LiChrosorb RP-18 Column with 1.44x10^{-2} M β-CD
Aqueous Solutions at Two Different pH

Compounds		pH 2.1	pH 6.8
		k´	
mandelic acid	+	4.88	0.75
	−	4.52	0.74
ortho-chloromandelic acid	+	119	11.3
	−	238	13.0
meta-chloromandelic acid	+	188	49.4
	−	245	56.6
para-chloromandelic acid	+	67	13.7
	−	67	13.7

Substantial enantioselectivity of β-CD complexation has
been found for ortho and meta chloromandelic acids as it
is shown in Figure 8 (21) and in Table II (21,34).
Chlorine - a substituent at the ortho or meta position
remarkably enhances enantioselectivity as compared to that
observed for mandelic acid itself, while at the para
position it reduces this enantioselectivity to undistin-
guishable values (see Table II).
The same procedure has been applied for resolution of
mephenytoin and some barbiturates into enantiomers (26).

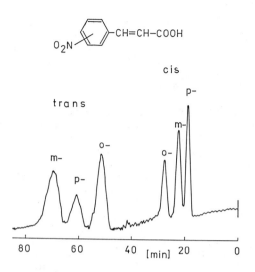

Figure 7. Chromatogram of cis-trans mixture of o-, m-, and p-nitrocinnamic acids performed on 10-μm LiChrosorb RP-18 column (100 × 4.6 mm i.d.) with 2.4 × 10^{-3} M β-CD. Solvent composition: 4 vol% methanol in an aqueous buffer of pH 4.2; flow rate: 1.5 mL/min; temperature: 25 °C.

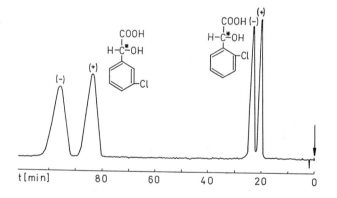

Figure 8. Chromatogram of a mixture of racemic o- and m-chloromandelic acids performed on 10-μm LiChrosorb RP-18 column (250 × 4.5 mm i.d.) with 14.4 x 10^{-3} M β-CD. Solvent composition: aqueous buffer of pH 6.8; flow rate: 1.2 mL/min; temperature: 25 °C. (Reprinted with permission from ref. 21. Copyright 1983 Elsevier Science Publishers.)

Figure 9 shows the examples of separations of racemic
mixtures of methylphenobarbital and mephenytoin performed under optimal conditions available (37). It has
been found that β -CD complexation results in a distinct enantioselectivity in the case of mephenytoin and
barbiturates which have a chiral center in the pyrimidine ring. The resolution of barbiturate enantiomers is
due to the different stabilities of their diastereoisomeric β -CD complexes, while the separation of
mephenytoin enantiomers results from the difference in
their adsorption on the RP phase. The latter case should
be considered further. It has been already suggested
(18) that the adsorption of CD complexes in which guest
molecules are entirely immersed in the CD cavity is low
on RP phases. The distinct adsorption arises from the
part of the molecule which is outside the cavity. Taking
into account this fact and the remarkable difference in
the adsorption of β -CD·mephenytoin diastereoisomers one
may conclude that a significant difference must exist
between immersion of mephenytoin enantiomers in the
β -CD cavity.
    It seems to also be worth mentioning that the descri-
bed procedure has been used for micro-preparative separations of mephenytoin and hexobarbital enantiomers (26)
β -CD solutions were also successfully used for resolution of 1-[2-(3-hydroxyphenyl)-1-phenylethyl]-4-(3-me-
thyl-2-butenyl) piperazine enantiomers in RP systems
(20). An especially interesting example of the application of γ -CD is the separation of optical isomers of
D,L - norgestrel (27).

Concluding Remarks.

In conclusion two facts should be noted. First, all the
chromatograms quoted above have been followed with UV
detectors. The usefulness of another detectors (e.g.
polarographic, refractive index) in CD solutions has
not yet been proved. Second, CD additions to mobile
phase solutions of RP systems are always followed by
a loss of column efficiency (ca 30%).This problem however demands more detailed studies.
    The method surveyed seems to be advantageous and
to some extent complementary to the important methods
in which the commercially available CD silica stationary phases are used. Moreover it involves sometimes additional factors which could improve separation such as
adsorption on an RP phase or second stage complexation.
The surveyed method has not been totally explored both
in theory and/or in practice. The newest results achieved via methylated CD's confirm this opinion(38,39).
Therefore still it is to early to draw general conclusions.

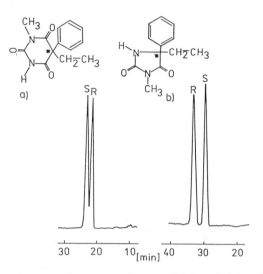

Figure 9. Chromatograms of racemic mixtures of (a) methylphenobarbital and (b) mephenytoin performed on 5-μm LiChrosorb RP-18 column (250 × 1 mm i.d.) with 2.2 × 10^{-3} M β-CD. Solvent composition: 20 vol% ethanol–aqueous buffer solution of pH 6.6 containing 0.5 vol% of diethylamine; flow rate: 30 μL/min; temperature: 25 °C.

Literature Cited

1. Szejtli, J. "Cyclodextrins and Their Inclusion Complexes", Akademiai Kiado; Budapest,1982;English.
2. Hinze, W.L.Sep.Purif.Methods 1981, 10, 159-237.
3. Smolkova-Keulemansova, E. J. Chromatogr. 1982, 251,
   17-34.
4. Fujimura, K.; Ueda, T.; Ando, T. Anal. Chem. 1983,
   55, 446-450,
5. Kawaguchi, Y.; Tanaka, M.; Nakae, M., Funazo, K.;
   Shono, T. Anal. Chem. 1983, 55, 1852-1857.
6. Tanaka, M.: Kawaguchi, Y.; Nakae, M.; Mizobuchi, Y.;
   Shono, T. J. Chromatogr. 1984, 299, 341-350.
7. Tanaka, M.; Kawaguchi, Y.; Shono, T.; Uebori, M.;
   Kuge, Y. J. Chromatogr. 1984, 301, 345-353.
8. Armstrong, D.W.; DeMond, W. J. Chromatogr. Sci.
   1984, 22, 411-415.
9. Armstrong, D.W. J. Liq. Chromatogr. 1984, 7 (s-2),
   353-376.
10. Armstrong, D.W.; DeMond, W.; Alak, A.; Hinze, W.L.;
    Riehl, T.E.; Bui, K.H. Anal. Chem. 1985, 57, 234-237.
11. Armstrong, D.W. U. S. Patent 4539399, Sept 3, 1985.
12. Hinze, W.L., Riehl, T.E,; Armstrong, D,W.; DeMond, W.;
    Alak, A.; Ward, T. Anal. Chem. 1985, 57, 237-242.
13. Armstrong, D.W.; DeMond, W.; Czech, B.P. Anal. Chem.
    1985, 57, 481-484.
14. Armstrong, D.W.; Alak, A.; DeMomd, W,; Hinze, W.L.;
    Riehl, T.E. J. Liq. Chromatogr. 1985, 8(2), 261-269.
15. Armstrong D.W.; Ward T.J.; Czech A.; Czech B.;
    Bartsch R.A. J. Org. Chem. 1985, 50, 5556-5559.
16. Ward, T.J.;Armstrong, D.W. J. Liq. Chromatogr. 1986,
    9, 407-423.
17. Dębowski, J.; Sybilska, D.; Jurczak, J. J. Chromatogr.
    1982, 237, 303-306.
18. Sybilska, D.; Lipkowski, J.; Wójcikowski, J.
    J. Chromatogr. 1982, 253, 95-100.
19. Dębowski, J.; Sybilska, D.; Jurczak, J. Chromatogra-
    phia 1982, 16, 198-200.
20. Nobuhara, Y.; Hirano, S.; Nakanishi, Y. J. Chromatogr.
    1983, 258, 276-279.
21. Dębowski, J.; Sybilska, D.; Jurczak, J. J. Chromatogr.
    1983, 282, 83-88.
22. Sybilska, D.; Dębowski, J.; Jurczak, J.; Żukowski, J.
    J. Chromatogr. 1984, 286, 163-170.
23. Żukowski, J.; Sybilska, D.; Jurczak, J. Anal. Chem.
    1985, 57, 2215-2219.
24. Dębowski, J.; Jurczak, J.; Sybilska, D.; Żukowski, J.
    J. Chromatogr. 1985, 322, 206-210.
25. Dębowski, J,; Grassini-Strazza, G.; Sybilska, D.
    J. Chromatogr. 1985, 349, 131-136.
26. Sybilska, D.; Żukowski, J.; Bojarski, J. J. Liq.
    Chromatogr. 1986, 9, 591-606.

27. Gazdag,M.; Szepesi,G.; Huszar,L.J.Chromatogr.1986, 351, 128-135.
28. Dębowski,J.; Sybilska,D. J.Chromatogr. 1986,353, 409-416.
29. Hinze, W.L.; Armstrong, D.W. Anal, Lett. 1980, 13, 1093-1104.
30. Armstrong, D.W.; Stine, G.Y. J. Am. Chem. Soc. 1983, 105, 2962-2964.
31. Uekama, K.; Hirayama, F.; Nasu, S.; Matsuo, N.; Irie T. Chem. Pharm. Bull. 1978, 26, 3477-3484.
32. Horvath, C.; Melander, W.; Nahum, A. J. Chromatogr. 1979, 186, 371-403.
33. Matsui, Y.; Mochida, K. Bull. Chem. Soc, Jpn. 1979, 52, 2808-2814.
34. Dębowski, J. Ph. D. Thesis, Polish Academy of Sciences, Warsaw, 1984.
35. Żukowski J.; Sybilska D. Institute of Physical Chemistry, Warsaw, unpublished data.
36. Dalgliesh, C.E. J. Chem. Soc. 1952, 3940.
37. Żukowski J.; Sybilska D. Institute of Physical Chemistry, Warsaw, unpublished data.
38. Tanaka, M.; Miki, T.; Shono, T. J, Chromatogr. 1985, 330, 253-261.
39. Żukowski, J.; Sybilska, D.; Bojarski, J.; J, Chromatogr. 1986, 364, 223-230.

RECEIVED May 11, 1987

Chapter 13

Least-Squares Iterations: Nonlinear Evaluation of Cyclodextrin Multiple Complex Formation with Static and Ionizable Solutes

Larry A. Spino and Daniel W. Armstrong

Department of Chemistry and Biochemistry, Texas Tech University, Lubbock, TX 79409-4260

Equations are derived which take into account the formation of cyclodextrin to substrate complexes other than simple one to one host:guest associations. An equation is also derived which describes the binding of a mono-protic species in which either its ionized or unionized form could bind to one or two cyclodextrin molecules. Because multiple binding constants are difficult to evaluate graphically, a non-linear least squares computer program is utilized. The approach works equally well for the determination of binding constants in micellar media.

A few recent reports have indicated that multiple cyclodextrin formation in aqueous solution is more common than once believed (1-8). Gelb et al. determined 1:2 (substrate:cyclodextrin {CD}) binding constant ratios but failed to obtain individual constant values (2). Connor et al. used potentiometric, spectrophotometric, solubility and competitive indicator methods to evaluate the binding constants of one substrate molecule bound to two CD molecules (5-7). Some of these methods gave substantial relative errors for the binding constants while other techniques were applicable to a limited number of compounds.

Armstrong et al. developed a chromatographic technique which could be used to evaluate the stoichiometry and all relevant binding constants for most substrate-CD systems (8). This method was not dependent on a solute's spectroscopic properties, conductivity, electrochemical behavior, or solubility. This work presented theory and chromatographic evidence for multiple cyclodextrin complex formation. Previous theoretical work considered only 1:1 complex formation (9-12). A two to one complexation equation was derived by expanding on the equation first used in 1981 to describe the 1:1 complexation behavior of a solute in a pseudophase system (13,14). Using this method, it was demonstrated that closely related compounds such as structural isomers of nitroaniline could exhibit different binding behaviors (8).

One form of the psuedophase retention equation is shown below (14). It relates LC retention (as the capacity factor, k') to the binding of a solute to cyclodextrin, K_1, and to the concentration of cyclodextrin in the mobile phase, [CD].

$$\frac{1}{k'} \text{ or } \frac{R_f}{1-R_f} = \frac{1}{\emptyset K[A]} + \frac{K_1[CD]}{\emptyset K[A]} \tag{1}$$

The terms in the denominator of the right side of Equation 1 include $\emptyset$, the phase ratio; A, stationary phase adsorption site; and K, the association constant of a solute to the stationary phase binding site. The binding constant, K_1, can be evaluated graphically (by plotting 1/k' versus [CD]) or by linear least squares. When complex equilibria are involved, Equation 1 deviates from linearity. In cases where two cyclodextrin molecules bind to a simple solute the correct pseudophase retention equation is (8):

$$\frac{1}{k'} \text{ or } \frac{R_f}{1-R_f} = \frac{1}{\emptyset K[A]} + \frac{K_1[CD]}{\emptyset K[A]} + \frac{K_1 K_2 [CD]^2}{\emptyset K[A]} \tag{2}$$

where K_2 is the second binding constant. Higher complexes contain additional analogous terms:

$$\frac{1}{k'} \text{ or } \frac{R_f}{1-R_f} = \frac{1}{\emptyset K[A]} + \frac{K_1[CD]}{\emptyset K[A]} + \frac{K_1 K_2 [CD]^2}{\emptyset K[A]} + \frac{K_1 K_2 K_3 [CD]^3}{\emptyset K[A]} + \cdots \tag{3}$$

A graphical solution for equation 2 is possible but somewhat complicated as some assumptions must be made (8). Graphical evaluation of more complex equilibria becomes very difficult if not impossible. As will be shown in this work, appropriate nonlinear least squares (NLLSQ) programs can be used to quickly evaluate and solve a variety of simple and complex equilibria problems. In addition to solving a number of the more difficult cyclodextrin complexation problems, data in the literature will be re-evaluated.

Experimental

All computations were performed on an Apple IIc personal computer. Values for the curves were generated using basic from an Apple DOS Master version 3.3. Graphs were drawn using Cricket Graph software for the Macintosh by Jim Rafferty and Rich Norling, Cricket Software, 3508 Market St., Suite 206, Philadephia, PA 19104. All nonlinear least squares (NLLSQ) iterations were performed with a NLLSQ program written for the Apple II+ version 1.2 by the CET Research Group, PO BX 2026, Norman, OK 73070, copyright 1981, 1983.

This particular NLLSQ program used the Marquardt method of expanding a model in a truncated Taylor's type series and solves for improved estimates of parameters in an iterative manner. The very nature of this strategy is that it finds a particular direction to move in search of better parameter estimates which is not exactly like the normal truncated Taylor's series nor that of the direction of "steepest descents".

Briefly put, the user introduces an equation into the body of the program. This equation is written in a form which takes into account independent and dependent variables. The user next incorporates this equation into a file to be recalled when needed at a later time. Data is next entered or taken from other compatible sources for later use. The NLLSQ file is loaded and the previously stored equation is then recalled. This particular program allows different programs to be stored and used when needed. This NLLSQ version is written so that the user is prompted to select needed items such as to make new array space, to enter the data file name and so on to start the iteration sequence. This program also requires that the user initiate guesses for the dependent variables, in this case, the binding constants. The program next cycles through the initiated guesses and when values are found by convergence, the results are listed and may be printed.

Results and Discussions

Curves generated by using equation 3 of 1/k' versus [CD] for 1:1 to 1:5 solute to CD complexes are shown in Figures 1 and 2. In Figure 1, binding constants were chosen so that the first binding constant was larger than the others. Figure 2 shows curves in which the product of the binding constants are equal, that is, $K_1K_2K_3K_4K_5 = K_1K_2K_3K_4 = K_1K_2K_3 = K_1K_2 = K_1$ for 1:5, 1:4, 1:3, 1:2 and 1:1 complexes respectively. Note that 1:1 complexes are linear as predicted by Equation 1 but higher complexes give curves of increasing slope. Visually inspecting plots of 1/k' versus [CD] makes distinguishing 1:2 complexes from 1:1 behavior difficult in this case. In many instances, complexes higher than 1:1 stoichiometries may also be indistinguishable from one another. These problems have been addressed previously (8). By using available NLLSQ programs, the difficulties of predicting stoichiometry by visually inspecting plots could be circumvented.

Table 1 lists values found by using a NLLSQ program to iterate for 1:1 and 1:2 stoichiometries using appropriate forms of Equation 3. As can be seen from Table 1, NLLSQ estimates for 1:2 solute:CD complexes were similar to those previously reported. Note that the 1:1 solute:CD binding constant for prostaglandin B_2 found by the NLLSQ program is close to the product of K_1K_2 for the NLLSQ value found for 1:2 solute:CD binding.

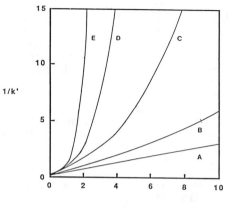

Figure 1. $1/k'$ versus $[CD]$ for complexes of 1:1 to 1:5
solute to cyclodextrin molecules. Binding
constants used are as follows:
       A, 1:1, $K_1=3000$
       B, 1:2, $K_1=300$, $K_2=10$
       C, 1:3, $K_1=300$, $K_2=50$, $K_3=5$
       D, 1:4, $K_1=100$, $K_2=30$, $K_3=5$, $K_4=0.5$
       E, 1:5, $K_1=200$, $K_2=100$, $K_3=30$, $K_4=5$, $K_5=1$
  $\emptyset K[A]=0.1$

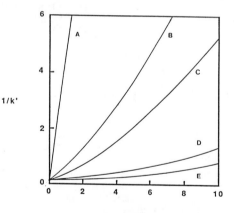

Figure 2. $1/k'$ versus $[CD]$ for complexes of 1:1 to 1:5
solute to CD molecules. The product of the
binding constants for each complex is equal. The
binding constants are as follows:
       A, 1:1, $K_1=5000$
       B, 1:2, $K_1=500$, $K_2=10$
       C, 1:3, $K_1=250$, $K_2=10$, $K_3=2$
       D, 1:4, $K_1=50$, $K_2=10$, $K_3=5$, $K_4=2$
       E, 1:5, $K_1=50$, $K_2=10$, $K_3=5$, $K_4=2$, $K_5=1$
  $\emptyset K[A]=0.1$

Table I. Binding constant values found by the NLLSQ program and compared to binding constant values previously reported

Compound	NLLSQ 1:1 K_1	Reported 1:1 K_1	NLLSQ 2:1 K_1	K_2	Reported (8) 2:1 K_1	K_2
prostaglandin B_1	----	970-1200 (16-17)	122	11.5	144	7.3
prostaglandin B_2	1447	709 (16)	284	6	95	6.2
4,4'-biphenol	----	----	135	47	300	102
o-nitroaniline	23	23	21	0.84	23	--
m-nitroaniline	73	73 (8)	73	0.03	73	--
p-nitronailine	----	----	--	----	430	32

Table II gives binding constants and partition coefficient values (P) calculated by the NLLSQ program using reported capacity factors for solutes bound to micelles. These values were determined on two different LC columns. One to one equations were used to obtain the binding constants (Equation 1). This table shows the ability of this particular NLLSQ routine to perform linear least square approximations. NLLSQ programs usually are less accurate at this. The mobile phase in this study was composed of solutions of sodium dodecylsufate (SDS). These binding constants were converted to partition coefficients by Equation 4.

$$K_{mw} = V(P_{mw} - 1) \qquad (4)$$

where K_{mw} is the binding constant of a solute between the micelle and the bulk water phase, V is the molar volume and P_{mw} is the partition coefficient of a solute between the micelle and the bulk water phase.

Although not known to the authors at present, cases of three to one or higher solute:CD complexation may be possible particularly when larger molecules are studied in CD systems. In such cases, it would be interesting to see how this specific NLLSQ program could handle the iterations and if the correct binding constants could be found. Table III gives these results and the number of cycles the program took to come to its conclusions when different guesses were first initiated for the dependent variables.

Figures 3 and 4 are plots of varying binding constants for 1:2 and 1:3 solute to cyclodextrin complexes using equations 2 and 3 respectively. In these plots, the product of the binding constants are equal to one another while the individual constants vary. It is apparent that the magnitude of the first binding constant (K_1) has the greatest effect on the relative position of the curves. This should not be surprising in view of the fact that K_1 appears in all of the terms of the right side of Equation 3 except the first.

Table II. Binding constants (K_{mw}) and partition coefficients (P_{mw}) using reported capacity factors [13] for solutes bound to micelles on two different columns (SDS mobile phase with C_{18} reversed stationary phase and SDS mobile phase with alkyl nitrile stationary phase)

Compound	C_{18} Reversed Phase Column		
	P_{mw} Reported	K_{mw} NLLSQ	P_{mw} Calculated
hydroquinone	14.5	3.81	16.32
resorcinol	27.1	7.9	32.78
p-nitrophenol	32.5	10.29	42.4
p-nitroaniline	72.3	11.83	48.59
	Alkyl Nitrile Column		
resorcinol	25.3	6.4	26.75
p-nitrophenol	32.9	8.02	33.3
p-nitroaniline	79.8	22.67	92.9

Table III 3 to 1 CD to substrate example. Data was generated with $K_1=100$, $K_2=10$, $K_3=2$, and $K[A]\emptyset=0.01$ from Equation 3. Twelve generated points were used

	Prompted Guess				NLLSQ Result				Cycles
	K_1	K_2	K_3	$K[A]\emptyset$	K_1	K_2	K_3	$K[A]\emptyset$	Required
1st guess	100	10	2	0.01	100.13	9.98	2.01	0.0099	6
2nd guess	100	100	100	100	100.13	9.98	2.01	0.0099	12
3rd guess	1000	1000	1000	1000	100.13	9.98	2.01	0.0099	15

pH Dependent Complexations

There are a variety of organic acids and bases that have protonated and unprotonated forms at different pH's. Each form of the solute can have a distinct binding constant to a cyclodextrin or micelle. Coupling these pH effects with multiple complexation behavior results in a somewhat complicated system. Solutes which have geometric cis-trans isomers as well as are pH dependent have been previously reported for 1:1 complexes (11). Sybilska et al. derived an equation which related the absorption of both neutral and anionic species on a reversed stationary phase (10) as shown in Equation 6.

$$t_{obs} = \frac{t_{HA} + t_{A^-}K_a/[H^+] + t_{HA-CD}K^0[CD] + t_{(A-CD^-)}K^-[CD]K_a/[H^+]}{1 + K_a[H^+] + K^0[CD] + K[CD]K_a/[H^+]} \quad (6)$$

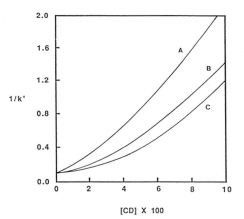

Figure 3.   1/k' versus [CD] plotting Equation 2.   The
product of the binding constants for curves A, B
and C are equal.   For curve A; K_1=100 and K_2=10,
for curve B; K_1=K_2=31.623 and for curve C; K_1=10
and K_2=100. $\emptyset K[A]$=0.1

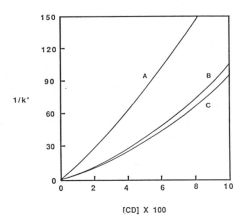

Figure 4.   1/k' versus [CD] plotting Equation 3.   For curve
A; K_1=1000, K_2=100 and K_3=10, for curve B;
K_1=K_2=K_3=100 and for curve C; K_1=10, K_2=100 and
K_3=1000. $\emptyset K[A]$=0.1

where t denotes retention times and the subscripts obs, HA, A⁻ HA-CD and (A-CD⁻) refer to overall (measured) values and to the retention of neutral, anionic, neutral-complexed and anionic-complexed species. K_a is the acidity constant and the remaining K's refer to the stability constants of 1:1 complexes of neutral and anionic species.

This work expanded on material previously reported by Uekama et al. relating observed retention times of ionic species and the concentration of cyclodextrin in the mobile phase (15).

$$t_{obs} = \frac{t'_o + t_c K_c [CD]_m}{1 + K_c [CD]_m} \qquad (7)$$

where t'_o is the retention time of the sorbate, t_c is the retention time of the sorbate-CD complex and K_c is the stability constant of the 1:1 complex.

Still, Sybilska et al. expanded the non-linear function $t_{obs} = t([H^+][CD])$ into a Taylor's series but neglected the nonlinear components. The parameters K^o, K^- and the individual capacity factors k'_{HA}, k'_{A^-}, $k'_{(HA-CD)}$ and $k'_{(A-CD^-)}$ were calculated (10). Neutral m-nitrobenzoic acid and anionic m-nitrobenzoate acid complexed to α-CD were found to have binding constants equal to 408± 21 and 486± 16 respectively. For p-nitrobenzoic acid K^o and K^- were found to be 473 ± 16 and 359 ± 16 respectively. Using the NLLSQ program would allow one to include these non-linear terms and give a more accurate picture of the binding of species which have neutral and ionic forms that can exist in a pseudophase.

It is possible that these ionic forms could exhibit 1:2 substrate:CD complexation, particularly para isomers, as shown for p-nitroaniline(8). Since Sybilska's working equations are non-linear for 1:1 complexes, graphing these equations would make calculating the binding constants difficult without first making some simplifing assumptions. This is where NLLSQ type programs would be useful. It would also be advantageous to have the equation in the form of capacity factors.

For a 1:2 substrate:CD complex in which the substrate is pH dependent, nine equilibria must be considered:

$$HA + S \overset{K}{\rightleftharpoons} HAS \qquad\qquad K = \frac{[HAS]}{[HA][S]} \qquad (7)$$

$$HA + CD \overset{K_1}{\rightleftharpoons} HACD \qquad\qquad K_1 = \frac{[HACD]}{[HA][CD]} \qquad (8)$$

$$\text{HACD} + \text{CD} \xrightleftharpoons{K_2} \text{HA(CD)}_2 \qquad , \; K_2 = \frac{[\text{HA(CD)}_2]}{[\text{HACD}][\text{CD}]} \qquad (9)$$

$$\text{HA} + 2\text{CD} \xrightleftharpoons{K_1 K_2} \text{HA(CD)}_2 \qquad , \; K_1 K_2 = \frac{[\text{HA(CD)}_2]}{[\text{HA}][\text{CD}]^2} \qquad (10)$$

$$\text{HA} + \text{H}_2\text{O} \xrightleftharpoons{K_a} \text{H}_3\text{O}^+ + \text{A}^- \qquad , \; K_a = \frac{[\text{H}^+][\text{A}^-]}{[\text{HA}]} \qquad (11)$$

$$\text{A}^- + \text{S} \xrightleftharpoons{K_o} \text{AS}^- \qquad , \; K_\circ = \frac{[\text{AS}^-]}{[\text{A}^-][\text{S}]} \qquad (12)$$

$$\text{A}^- + \text{CD} \xrightleftharpoons{K_{1a}} \text{ACD}^- \qquad , \; K_{1a} = \frac{[\text{ACD}^-]}{[\text{A}^-][\text{CD}]} \qquad (13)$$

$$\text{ACD}^- + \text{CD} \xrightleftharpoons{K_{2a}} \text{A(CD)}_2^- \qquad , \; K_{2a} = \frac{[\text{A(CD)}_2^-]}{[\text{ACD}^-][\text{CD}]} \qquad (14)$$

$$\text{A}^- + 2\text{CD} \xrightleftharpoons{K_{1a} K_{2a}} \text{A(CD)}_2^- \qquad , \; K_{1a} K_{2a} = \frac{[\text{A(CD)}_2^-]}{[\text{A}^-][\text{CD}]^2} \qquad (15)$$

where HA is the free protic solute, S is the stationary adsorption site, CD is a cyclodextrin molecule, HACD is a 1:1 solute:CD complex, HA(CD)_2 is a 1:2 solute:CD complex, A^- is the anion of the free protic solute, ACD^- is a 1:1 anionic solute:CD complex, A(CD)_2^- is a 1:2 anionic solute:CD complex and the respective equilibrium constants are K, K_1, K_2, K_3, K_a, K_{1a} and K_{2a}. The total amount of solute HA_t is given by:

$$\text{HA}_t = \text{HA} + \text{HAS} + \text{HACD} + \text{HA(CD)}_2 + \text{A}^- + \text{AS}^- + \text{A(CD)}^- + \text{A(CD)}_2^- \qquad (16)$$

Substituting into the standard chromatographic definition of capacity factor (Equation 17), one obtains Equation 18.

$$k' = \frac{\text{Amount of solute in the stationary phase}}{\text{Amount of solute in the mobile phase}} \qquad (17)$$

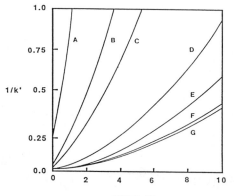

Figure 5.   1/k' versus [CD] plotting Equation 18. For curve
             A; pH=8, for curve B; pH=7.3, for curve C; pH=7,
             for curve D; pH=6.3, for curve E; pH=6 and 5.8,
             for curve F; pH=5 and for curve G; pH=4.3 and
             4.0.

$$k' = \frac{\emptyset([HAS] + [AS^-])}{[HA] + [HACD] + [HA(CD)_2] + [A^-] + [ACD^-] + [A(CD)_2^-]} \quad (18)$$

Substituting [HAS], [AS⁻], [HACD], [HA(CD)$_2$], [A⁻], [ACD⁻] and [A(CD)$_2$⁻] using equations 7-15, cancelling [HA], letting X = $\emptyset[S](K_oK_oK_a\{1/[H^+]\})$ and rearranging gives an equation which describes the LC retention behavior of monoprotic solutes which complex two cyclodextrin molecules.

$$\frac{1}{k'} = \frac{1}{X} + \frac{K_1[CD]}{X} + \frac{K_1K_2[CD]^2}{X} + \frac{K_a}{[H^+]X} + \frac{K_{1a}K_a[CD]}{[H^+]X} + \frac{K_{1a}K_{2a}K_a[CD]^2}{[H^+]X} \quad (19)$$

If K_2 and K_{2a} are equal to zero, then this equation reduces to a 1:1 complexation equation.

Figure 5 is a plot of 1/k' versus [CD] in Equation 19 for different pH's of an imaginary mono-protic acid that could bind to one or two cyclodextrin molecules in either protonated or deprotonated form. Realistic binding constants were used and a pK$_a$ of 5.82 was chosen. From the plot, we can see that if the binding constants of the prontonated and deprotonated forms are similar, the magnitude of the curve will still be larger at higher pH's because of the terms with [H⁺] in the denominator. Of course, the curvature of these slopes is dependent on the size of the binding constants.

By using a NLLSQ program and letting 1/k', [CD], [H⁺], and K$_a$ be the independent variables, the computer could iterate for the dependent variables, X, K_1, K_2, K_{1a} and K_{2a}. This would seem a logical alternative to extracting data from graphs which typically would be non-linear and difficult to interpret. By using a NLLSQ program, simplifying assumptions no longer would be necessary for the reason of only producing linear plots. Also, the problem of intercepts being inaccurate or close to zero for these linear simplifications would be circumvented.

Acknowledgment

The support of this work by the Department of Energy, Office of Basic Energy Science (DE-AS0584ER13159), is gratefully acknowledged.

Literature Cited

1. Harata, K.; Bull. Chem. Soc. Jpn. **1976**, 49, 1493.
2. Gelb, R. I.; Schwartz, L. M.; Laufer, D. A. J. Am. Chem. Soc. **1979**, 100, 3553.
3. Gelb, R. I.; Schwartz, L. M.; Laufer, D. A. J. Am. Chem. Soc. **1979**, 101, 1869.
4. Connors, K. A.; Rosanske, T. W. J. Pharm. Sci. **1980**, 69, 173.
5. Wong, A. B.; Lin, S. F.; Connors, K. A.; J. Pharm. Sci. **1983**, 72, 388.

6.   Connors, K. A.; Pendergast, D. D. J. Am. Chem. Soc. **1984**, 106, 7607.

7.   Pendergast, D. D.; Connors, K. A.; J. Pharm. Sci. **1984**, 73, 1779.

8.   Armstrong, D. W.; Nome, F.; Spino, L. A.; Golden, T. D. J. Am. Chem. Soc. **1986**, 108, 1418.

9.   Armstrong, D. W.; Stine, G. Y. J. Am. Chem. Soc. **1983**, 105, 2962.

10. Sybilska, D.; Lipkowski, J.; Woycikowski, J. J. Chromatogr. **1982**, 253, 95.

11. Sybilsa, D.; Debowski, J.; Jurezak, J.; Zukowski, J. J. Chromatogr. **1984**, 286, 163.

12. Arunyanart, M.; Cline-Love, L. J. Anal. Chem. **1984**, 56, 1557.

13. Armstrong, D. W.; Nome, F. Anal. Chem. **1981**, 53, 1662.

14. Armstrong, D. W. Sep. Purif. Methods **1985**, 14, 213.

15. Uekama, K.; Hirayama, F.; Nasu, S.; Matsuo, N.; Irie, T. Chem. Pharm. Bull. **1978**, 26, 3477.

16. Uekama, K.; Hirayama, F.; Ikeda, K.; Inaba, K. J. Pharm. Sci. **1977**, 66, 706.

17. Uekama, K.; Hirayama, F.; Irie, T.; Chem. Lett. (Japan) **1978**, 66.

RECEIVED January 20, 1987

Chapter 14

Gas Chromatographic Separation of Structural Isomers on Cyclodextrin and Liquid Crystal Stationary Phases

Eva Smolková-Keulemansová[1] and Ladislav Soják[2]

[1]Department of Analytical Chemistry, Charles University, 128 40 Prague 2, Albertov 2030, Czechoslovakia
[2]Chemical Institute, Comenius University, 842 15 Bratislava, Mlynská dolina, Czechoslovakia

To improve the effectiveness of the chromatographic separation, a comparison study has been carried out on cyclodextrin and liquid crystal stationary phases. Both materials function as "ordered" media: with cyclodextrins the inclusion complex formation predominates, whereas the liquid crystals enable interaction of compounds with the ordered structure of the mesophase. The properties of cyclodextrins were studied using a packed column, whereas the effect of the liquid crystals were enhanced by employing a capillary column. The stereoselective properties of these materials as stationary phases were studied with a set of alkylbenzoderivatives. The mechanism of the separation is discussed on the basis of the retention data obtained. The advantages and drawbacks of these phases are compared with conventional GC stationary phases and analytical applications are discussed.

Increasing requirements on analyses of isomeric compounds and the problems encountered in their separation necessitate a study of more efficient systems which exhibit a high selectivity. In gas chromatography new, selective stationary phases are studied. Attention is also focused on the use of substances with oriented molecules that permit selective separations; these properties are exhibited by e.g. inclusion compounds and liquid crystals.
  Although the interaction mechanisms are different with liquid crystals and cyclodextrins, their stereoselective properties are so important that it is desirable to deal with their potential use in gas chromatogra-

phy. The interaction mechanism with inclusion compounds
is based on a specific interaction during which a molecule (the guest) is inserted, the whole molecule or part
of it, into a cavity in another molecule (the host) in
order to attain a state with a minimum energy. Attention
has recently been centered on the utilization of the formation of inclusion compounds of cyclodextrins (CD)(1-4).
Cyclodextrins and their derivatives have several advantages over other hosts, such as the ability to form complexes both in the solid state and in solutions (5). The
ability of CD's to interact in the solid and liquid sta-
tes with liquid and gaseous substances has permitted
their use in liquid and gas chromatography. In liquid
chromatography, CD's are used as polymers or chemically
bonded stationary phases, or as the selective component
of the mobile phase in reversed-phase systems (4).

In gas chromatography, the selective properties of
these substances are used in GSC or GLC, where the substances act as a selective component of the stationary
liquid phase (4).

Wide applicability of CD's follows from the fact
that they occur as 6 to 8-membered rings(α-,β- and γ-CD)
with cavities of various sizes. The stability of the CD
complexes is primarily determined by the steric arrangement (i.e. the sizes and shapes of the guest molecules
and the CD cavity) and can be increased by hydrogen bonding. These factors form the basis for stereoselective
separations of substances based on their (CD's) use in
separation science.

Liquid crystals represent a transition between solid crystalline substances and isotropic liquids. On heating, mesophases are formed that have ordered structures
which can be nematic, smectic or cholesteric. On further
heating, the orientation is disturbed and the phases are
converted into an isotropic liquid. The long structure
of liquid crystals causes isomers with more drawn-out
shapes to be readily dissolved in the ordered liquid
crystal substrate (mesophase) thus yielding stronger sor-
bat-sorbent interactions.

Kelker (6) first used liquid crystals as stereospecific stationary phases in gas chromatography. Since then,
a great deal of attention has been paid to the separation
properties of this relatively wide group of substances
(7,8), used mainly as stationary phases in packed columns.
The present knowledge indicates that the sorbate shape
contributes relatively little to the overall retention
of the sorbate. However, it is possible to separate isomeric compounds with similar anisotropic properties that
cannot be commonly separated even on highly efficient columns ($n_{reg} = 10^6$ to 10^7 plates). To enhance the stereospecific properties of liquid crystals, it is necessary to
work in systems combining the structural selectivity of
the substances with the high efficiency of capillary columns, i.e. liquid crystals as stationary phases in
capillary gas chromatography (9-11).

The general relationship between the chromatographic selectivity and the liquid crystal structure have not yet been unambiguosly clarified. This problem has been studied in greater detail on 4,4-dialkoxyazoxybenzenes.

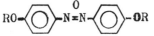

It has been found that the terminal groups of azoxybenzenes have a great effect on selectivity α, expressed for para-/meta-xylene. Replacement of the central azoxy group by the azo group leads to a strong decrease in the selectivity, even with the most selective arrangement of the terminal groups. The highest selectivity is exhibited by the asymetrical, short-chain derivative, methoxyethoxy-azoxybenzene (MEAB).

In this paper, we present experimental data on the above types of stationary phases, in order to evaluate the stereoselective properties of these phases and the possibilities of separating isomeric substances that are otherwise difficult to separate by current analytical procedures.

Experimental

The α-, β- and γ-CD preparations were obtained from Chinoin (Budapest, Hungary). The stationary phases were prepared by depositing CD from a dimetylformamide solution onto Chromosorb W 60-80 mesh . The solutions were 7 to 10% in CD which ensured complete coverage of the Chromosorb with cyclodextrin. The specific surface areas of the resulting stationary phases, measured by the thermal desorption method, were 1.4-2.0 $m^2.g^{-1}$; hence the effect of the surface are need not be considered in the treatment of the retention data. The glass columns used were 120 cm long and 2 to 3 mm i.d.

The liquid crystals were deposited in glass capillary columns up to 100 m long, 0.25 mm i.d. The columns were whet using the dynamic method, by flowing 1.5 to 20% liquid crystal solutions in chloroform at a flow rate of 1 $cm.s^{-1}$ through the column. The efficiency of the prepared columns was compared with that of common stationary phases (columns 100 m long and 0.25 mm i.d. exhibited an efficiency of up to 350 000 theoretical plates for hydrocarbons with $k \approx 5$).

The measurements with the CD's were performed on a Chrom 4 gas chromatograph with the flame-ionization detection (Laboratorní Přístroje, Prague, Czechoslovakia). The measurements with the liquid crystal phases were performed on a Perkin Elmer F-11 instrument with a flame-ionization detector.

With the CD phases saturated vapours of the test substances were injected with a Hamilton microsyringe,

whereas with the liquid crystals, liquid samples were
injected using a splitter.

Results and discussion

Retention of isomers. We have examined the effect of
alkyl groups on the retention behaviour, in dependence
on the chain length, branching and the relative positions in di- and trisubstituted benzene derivatives. The
experimental data are summarized in Table I.

Table I. Relative retention ($r_{2,1}$) of aromatic hydrocarbons on squalane (SQ), cyclodextrins (CD), and liquid
crystals (EBO, MEAB) as stationary phases

Hydrocarbon	Boiling point, °C	Relative retention, $r_{2,1}$				
		SQ^{80}	$\alpha\text{-}CD^{90}$	$\beta\text{-}CD^{90}$	EBO^{40}	$MEAB^{95}$
Benzene	80.1	1.00	1.00	1.00	1.00	1.00
Toluene	110.6	2.55	2.25	0.96	2.77	2.10
Ethylbenzene	136.2	5.38	8.64	1.50	5.92	3.48
p-Xylene	138.4	6.03	13.5	1.16	7.94	4.67
m-Xylene	139.1	6.15	1.52	0.94	7.22	4.14
o-Xylene	144.4	7.20	0.88	1.47	9.40	5.17
Isopropylbenzene	152.4	8.78	1.03	2.54	8.77	4.38
Propylbenzene	159.2	11.13	18.5	3.11	13.01	7.31

It is evident from the data for monoalkylbenzenes on
α-CD that the retention of n-alkylbenzenes is affected
by the high stabilities of the complexes formed, caused
by the location of the alkyl group in the cyclodextrin
cavity. Branching of the side chain leads to a pronounced decrease in the retention. These results are similar
those obtained for the interactions of n-alkanes and
branched alkanes with α-CD (12,13).
    It can be concluded from the retention data on β-CD
that the chain length and, to a certain extent, the degree of branching do not affect the inclusion process.
This suggests that the same part of the guest molecules
interact with β-CD cavity, i.e. the benzene ring is
oriented into the cyclodextrin cavity.
    Steric properties play an especially important role
with disubstituted benzene derivatives, for which the
typical elution order on α-CD is: o-, m-, p-isomer. The
greater retention of p-isomers is due to the close contact of the α-CD cavity with one of the substituents,
probably with the more hydrophobic one. The steric hindrance caused by substituents in the m- and especially
the o-position prevents penetration of the guest molecule into the α-CD cavity.
    Dialkylbenzenes are eluted in the same order on
β-CD as on α-CD (14). Xylenes are exceptional,

as the o-dimethyl isomer may at least partially enter the larger β-CD cavity, so that the elution order becomes: m-, p-, o-isomer. This fact is in agreement with the results of a comparison of a model of o-xylene with the size of the β-CD cavity. However, steric hindrance predominates with guests possessing more voluminous substituents. Thus o-isomer of the higher homologues elute first (Table II). Their separation from the m-isomers depends considerably on the relative size of the two substituents. Relatively small differences in the retention of the o- and m-isomers have been found for the ethyl- and propyltoluenes. On the other hand, weak interactions of 1-methyl-2-isopropylbenzene (o-cymene) and 1,2-diethylbenzene with the β-CD cavity were observed, owing to the more voluminous substituent in the ortho position. This results in greater differences in the relative retentions of these isomers. In all these cases, the p-substituted derivatives exhibit greatest degree of interaction and have the highest retention.

Table II. Retention data $(t_R^{\bullet})$ of benzene dialkyl derivatives on β-CD and γ-CD stationary phases at 90°C

Hydrocarbon	Boiling point, °C	$t_R^{\bullet a}$, s	
		β-CD	γ-CD
p-Xylene	138.4	22.0	24.0
m-Xylene	139.1	19.7	30.9
o-Xylene	144.4	24.1	79.5
m-Ethyltoluene	161.3	30.9	45.7
p-Ethyltoluene	162.0	43.8	41.4
o-Ethyltoluene	165.2	29.8	91.7
m-Diethylbenzene	181.5	57.5	81.4
p-Diethylbenzene	183.8	142.8	79.9
o-Diethylbenzene	184.2	38.2	153.7
m-Cymene	176.0	76.1	-
p-Cymene	177.1	110.3	77.2
o-Cymene	178.2	37.5	139.5
m-Propyltoluene	182.0	81.6	89.0
p-Propyltoluene	183.0	119.6	95.2
o-Propyltoluene	185.0	66.5	155.8

$^a t_R^{\bullet}$=corrected retention time = $t_R - t_M$

The experimental results with γ-CD indicate (Table II) that the retention of substances examinated is not perceptibly affected by inclusion due to the large cavity of the cyclodextrin (14). Therefore, the retention data for all the p-isomers of dialkylbenzenes are lowest in contrast to the highest values for these derivatives on β-CD. Among the disubstituted alkylbenzenes the o-isomers are retained most, owing to their more voluminous special arrangement. The weak interactions of the p- and m-isomers are reflected in the negligible differences in the retention. Consequently, their separation is virtuably imposible.

A general picture of the specific interactions of aromatics on α-, β- and γ-CD can be obtained by comparing the results of the chromatographic study with previously published data. The thermodynamic quantities indicate that only part of the benzene molecule is included in the α-CD cavity, whereas the contact with the β-CD cavity is very intimate. The published values of the formation equilibrium constants of the complexes formed also follows the order β- $\gg$ α- $>$ γ-CD for the compounds studied.

With liquid crystals, the differences in the longitudinal dimension of molecules and in their planarity play a major role in determining the strengh of interaction with guest molecules. Sterespecific effects are, in general, less pronounced than those observed with cyclodextrins. Therefore, in order to be able to use liquid crystals for separations, the modest liquid crystal selectivity has been combined with the high efficiency of capillary columns.

Among the alkylbenzenes, the stereospecific effect of liquid crystals is most marked in the separation of m- and p-xylene. The more linear p-xylene exhibiting a greater retention than the m-isomer (similar retention observed on α- and β-CD). The retention order on the liquid crystals, m-$<$ p-$<$ o-xylene is different from that on common polar and nonpolar stationary phases (p-$<$ m-$<$ o-isomer). In contrast to liquid crystals, o-xylene exhibited the lowest retention on α-CD.

The retention order of propyl- and isopropylbenzenes is the same on liquid crystals as on cyclodextrins. However, as can be seen from Table I, the difference in the relative retentions is exceptionally large on α-CD, which again illustrates the more pronounced stereospecific effects of cyclodextrins.

In order to compare and contrast the selectivities of the various stationary phases materials, the selectivity were calculated for the p/m xylene couple and are given in Table III. The high selectivity of CD and of other types of inclusion compounds is apparent from the data in this Table. Also, it is obvious that the liquid crystals also yield better selectivities than do the common stationary phases. This fact is further illustrated by a comparison of the dependence of α on the number of C atoms in the guest molecules of n-alkyl- and o-dialkylbenzenes, C_{14} to C_{17}, on a Carbowax 20M polar phase and on the liquid crystal (MEAB)(see Figure 1). It follows from the Figure that the MEAB selectivity for the separation of alkylbenzene isomers increases with increasing length of the longer alkyl chain of the dialkylbenzenes. The order of the selectivity factors is:

n- > methyl- > ethyl- > propyl- > butyl- > pentyl-2-alkylben-
zene, which is in the same order as the growth of their
retention. Therefore, the retention interval for the positional isomers of alkylbenzenes on MEAB is 50% greater
than on Carbowax 20M. Thus the separation of these substances is much easier.

Table III. Selectivity factor (of para-/meta-xylene) on
           columns with various packings

Stationary phase	Column temperature, $^\circ$C	Selectivity factor, α
Squalane	80	0.98
Carbowax 20M	80	0.96
EBO	40	1.10
MEAB	90	1.13
Benton 34[a] with a silicone phase	70	1.26
Ni(NCS)$_2$(4-methylpiridine)$_4$[a]	80	2.60
3.5% α-CD[a]	100	3.80

[a]Separation based on inclusion complex formation

     While the α dependence on the number of C atoms is
monotonic for Carbowax, with MEAB the α values alternate
with an increase in retention exhibited by n-alkylbenzenes having an odd numbers of C atom in the molecule and
by dialkylbenzenes that have an odd number of C atoms in
their longer alkyl chain (which is in agreement with a
greater linearity (length) of these structures). Therefore, liquid crystals retain more selectively 1-propyl-
2-pentylbenzene than 1,2-dibutylbenzene, compared with
common phases, which has a positive effect on their separation.
     Phenylalkanes (C_{10}-C_{13}) behave analogously and the
selectivity increases with a shift of phenyl toward the
end of the alkane chain and the elution order follows
the same trend as just described. The values also alternate, i.e. the phenylalkanes with odd numbers of C atoms
or with a phenyl on an odd carbon, exhibit an increased
retention. From the point of view of the separation of
positional isomers, the favorable effect of the increase
in the selectivity with the shift of the phenyl toward
the end of the carbon chain is partly compensated by the
retention alternation. For example, the selectivity factor α for the critical pair 6-phenyltridecane/7-phenyl-
tridecane on MEAB (1.015) is only slighthly higher as on
OV-101 (1.007).
     The selectivity of liquid crystals can also play a
role in the separation of linear n-alkene isomers. On
liquid crystal phases, alternation is pronounced in separations of alkenes with a double bond in the middle of

the hydrocarbon chain. Such hydrocarbons are difficult to separate on common stationary phases requiring $n_{reg} = 10^6$ to 10^7 plates due to fact that these isomers have very similar physico-chemical properties. Therefore, even a small difference in the linear dimension of the molecules, caused by the different positions of the double bond which results in changes in the zig-zag arrangement of the end of the carbon chain in or out of the molecule axis, gives rise to an altered elution order (Figure 2). This effect can be considered as quite a specific separation property of liquid crystals which facilitates the most difficult separation of these types of isomers.

Analytical applications. The differences in the stabilities of the complexes with α-, β- and γ-CD can be used to advantage for the separation of various disubstituted isomers of alkylbenzenes. An example is the separation of the isomers of diethylbenzene, methylpropylbenzenes (cymenes) and trimethylbenzenes on β-CD (refer to Figure 3).

The selectivity of liquid crystals, combined with the high efficiency of capillary columns, makes possible the separation of multicomponent mixtures, including critical pairs of isomers. This is illustrated by the separation of alkylbenzenes up to C_8 (Figure 4), alkylbenzenes C_{14} to C_{17} (Figure 5), and phenylalkanes C_{10} to C_{13} (Figure 6). In most cases, very rapid analyses (see Figure 3,4) or a better separation of multicomponent mixtures (see Figure 5,6) results from the use of liquid crystalline stationary phases.

Conclusions

The classical types of interactions that play a role in separation using conventional polar and nonpolar stationary phases yield limited possibilities for separation of isomers with similar properties. For a solution of this problem, a high degree of selectivity in the separation process is required. It has been shows that these requirements are met by both cyclodextrins and liquid crystals, used as stationary phases in gas chromatography.

An advantage of cyclodextrins over the common stationary phases is the high selectivity toward the isomeric substances. It has been demonstrated that many positional and geometric isomers can be separated by packed-column GSC in a very short time (i.e. analysis time does not exceed 2 minute) in separation of a mixture of o-, m- and p-isomers. From the analytical viewpoint, the low efficiency of the columns used is a disadvantage. Also, there are other drawbacks of the gas-solid chromatography using CD's: nonlinearity of the separation isotherm over a wider concentration range and poor reproducibility in the preparation of the CD columns utilized.

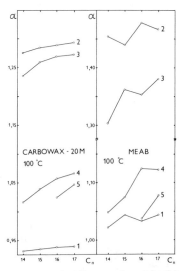

Figure 1. Dependence of the selectivity factor (α) for C_{14}-C_{17} alkylbenzenes on the C-number of homologous on column with Carbowax 20M and MEAB; l=n-alkyl-/l-methyl-2-alkyl-, 2= l-methyl-2-alkyl-/l-ethyl-2-alkyl-, 3= l-ethyl-2-alkyl-/l-propyl-2-alkyl-, 4=l-propyl-2-alkyl-/l-butyl-2-alkyl-, 5=l-butyl-2-alkyl-/l-pentyl-2-alkylbenzene.

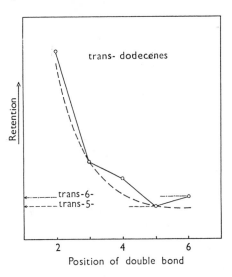

Figure 2. Dependence of the retention of isomeric trans-dodecenes on the position of double bond on a liquid crystal column.

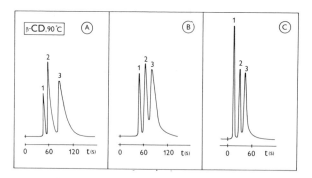

Figure 3. Separations of isomeric alkylbenzenes on
β-CD at 90°C; A=o-, m- and p-diethylbenzene, B=o-, m-
and p-methylisopropylbenzene, C=1,3,5-, 1,2,3- and
1,2,4-trimethylbenzene.

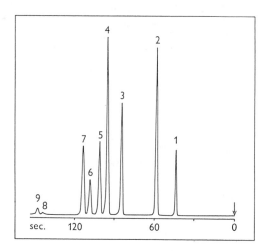

Figure 4. Separation of C_6-C_9 aromatic hydrocarbons
on the column with liquid crystal EBO, 48 m x 0.25 mm
at 40°C; 1=benzene, 2=toluene, 3=ethylbenzene, 4=m-
xylene, 5=p-xylene, 6=isopropylbenzene, 7=o-xylene,
8=n-propylbenzene and 9=styrene.

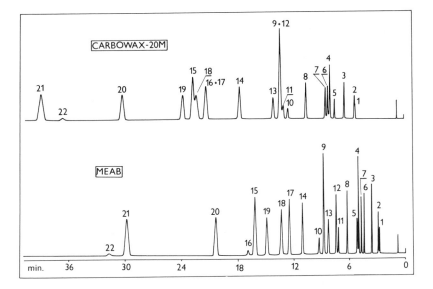

Figure 5. Separation of $C_{14}-C_{17}$ alkylbenzenes and
o-dialkylbenzenes obtained from dehydrogenation of
n-alkanes in columns with Carbowax 20M and MEAB;
1=1-butyl-2-butylbenzene, 2=1-propyl-2-pentylbenzene,
3=i-ethyl-2-hexylbenzene, 4=1-methyl-2-heptylbenzene,
5=n-octylbenzene, 6=1-butyl-2-pentylbenzene, 7=1-pro-
pyl-2-hexylbenzene, 8=1-etyl-2-heptylbenzene, 9=1-me-
thyl-2-oktylbenzene, 10=n-nonylbenzene, 11=1-pentyl-
2-pentylbenzene, 12=1-butyl-2-hexylbenzene, 13=1-pro-
pyl-2-heptylbenzene, 14=1-ethyl-2-octylbenzene, 15=1-
methyl-2-nonylbenzene, 16=n-decylbenzene, 17=1-pentyl-
2-hexylbenzene, 18= 1-butyl-2-heptylbenzene, 19=1-pro-
pyl-2-octylbenzene, 20= 1-ethyl-2-nonylbenzene, 21=1-
methyl-2-decylbenzene, 22=n-undecylbenzene.

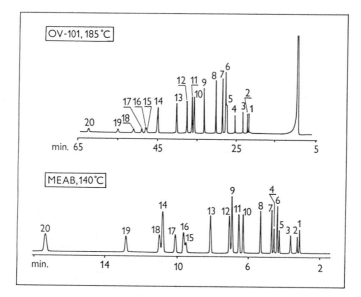

Figure 6. Separation of C$_{10}$-C$_{13}$ phenylalkanes in columns
with OV-101 and MEAB; 1=phenyldecane, 2=4-phenyldecane,
3=3-phenyldecane, 4=2-phenyldecane, 5=6-phenylundecane,
6=5-phenylundecane, 7=4-phenylundecane, 8=3-phenylun-
decane, 9=2-phenylundecane, 10=6-phenyldodecane, 11=5-
phenyldodecane, 12=4-phenyldodecane, 13=3-phenyldodeca-
ne, 14=2-phenyldodecane, 15=7-phenyltridecane, 16=6-
phenyltridecane, 17=5-phenyltridecane, 18=4-phenyltri-
decane, 19=3-phenyltridecane, 20=2-phenyltridecane.

On the other hand, capillary gas chromatography with
liquid crystals yields very good analytical separations
(even for critical pairs of isomers) as the GLC system
is used and an inherent lower selectivity compared to cyclodextrins is compensated for by the higher efficiency
of the capillary columns. Therefore, future work should
be directed toward reproducible preparation of capillary
columns with cyclodextrins and to other liquid crystals
with a higher selectivity.

Acknowledgments

We would like to thank Prof. J. Szejtli (Chinoin, Budapest, Hungary) for providing cyclodextrins and Prof. M.S.
Vigdergauz (Kuybishev, U.S.S.R.)and Dr. G. Kraus (Halle,
G.D.R.)for providing liquid crystals, which have made
this work possible.

Literature Cited

1. Hinze, W.L. Sep. Purif. Methods 1981, 10, 159.
2. Smolková-Keulemansová, E. J. Chromatogr. 1982, 251,
   17.
3. Sybilska, D.; Smolková-Keulemansová, E. In "Inclusion
   Compounds"; Davies, E., Ed.; Academic Press: New York,
   1984; Vol. III, p. 173-243.
4. Smolková-Keulemansová, E. In "Cyclodextrins and their
   uses in various industries"; Duchéne, D., Ed.; Paris,
   in press.
5. Szejtli, J. In "Cyclodextrins and Their Inclusion
   Complexes"; Akademiai Kiado: Budapest, 1982.
6. Kelker, H. Ber. Bunsenges. phys. Chem. 1963, 67, 698.
7. Vigdergauz, M.S.; Vigalok, R.V.; Dimitrieva, G.V.
   Usp. Khimii 1981, 50, 1943.
8. Witkievicz, Z. J. Chromatogr. 1982, 251, 311.
9. Soják, L.; Kraus, G.; Farkaš, P.; Ostrovský, I.
   J. Chromatogr. 1982, 249, 29.
10. Soják, L.; Kraus, G.; Ostrovský, I. J. Chromatogr.
    1985, 323, 417.
11. Soják, L.; Ostrovský, I.; Farkaš, P.; Janák, J.
    J. Chromatogr. 1986, 356, 104.
12. Mráz, J.; Feltl, L.; Smolková-Keulemansová, E.
    J. Chromatogr. 1984, 286, 17.
13. Smolková-Keulemansová, E.; Krýsl, S.; Feltl, L.
    J. Incl. Phenomena 1985, 3, 183.
14. Smolková-Keulemansová, E.; Neumannová, E.; Feltl, L.
    J. Chromatogr. in press.

RECEIVED December 19, 1986

Chapter 15

High-Performance Liquid Chromatography Using a β-Cyclodextrin-Bonded Silica Column: Effect of Temperature on Retention

Haleem J. Issaq, Maureen L. Glennon, Donna E. Weiss, and Stephen D. Fox

Program Resources, Frederick Cancer Research Facility, National Cancer Institute, P.O. Box B, Frederick, MD 21701

The effect of temperature, and of temperature and pH on the retention of a selected group of compounds using a beta-cyclodextrin column was studied. The results indicated that a plot of lnk' vs. 1/T gave linear relationships for anthraquinone, methyl anthraquinone, ethyl anthraquinone, naphthalene and biphenyl using a mobile phase of methanol/water. However, a non linear relationship was observed for a selected group of dipeptides employing a mobile phase of methanol/ammonium acetate at the following pH's: 4, 5.5 and 7. The retention times decreased with an increase in the temperature of the column except that for certain dipeptides the retention times increased. The separation factor (α) values decreased by approximately 10% with increase in column temperature from 25°C to 77°C.

The use of high performance liquid chromatography (HPLC) for the separation of various groups of compounds, using a β-cyclodextrin bonded silica column is well established (1-8). The effect of the volume of organic modifier in the eluent was studied (9), no quantitative data has been published on the effect of temperature on retention using a beta-cyclodextrin bonded silica. This study deals with the effect of temperature on the capacity factor of (a) naphthalene and biphenyl, (b) anthraquinone, methyl- and ethyl anthraquinone, and (c) p-nitroaniline. Also, the effect of temperature and pH on the retention and resolution of a selected group of dipeptides was investigated.

EXPERIMENTAL

The HPLC system consisted of a Hewlett Packard (Avondale, PA, USA) HP1090 Liquid Chromatograph equipped with a heater, a photo diode

array detector, an HP 85B system controller, and an HP 3392A
integrator. The mobile phase solvents were filtered using a
Millipore filter holder (Milford, MA, USA) with Millipore filters
having a pore size of 0.5 um organic and 0.45 um aqueous. The pH
meter used to adjust the pH of the mobile phases was a Fisher Accumet
Model 320 (Fair Lawn, NJ, USA). The samples were dissolved in
methanol or 0.1 HCl (dipeptides). Separations were carrried out on a
250 mm x 4.6 mm Cyclobond I column from Advanced Separation
Technologies, Inc. (Whippany, NJ, USA) which has beta-cyclodextrin
bonded to 5u spherical silica particles.

MATERIALS. Anthraquinone, methyl- and ethylanthraquinone,
naphthalene and biphenyl were purchased from Aldrich Chemical Co.,
Inc. (Milwaukee, WI, USA). Dipeptides were obtained from Sigma (St.
Louis, MO, USA), Chemical Dynamics Corp. (South Plainfield, NJ, USA),
and U.S. Biochemical Corp. (Cleveland, OH, USA). The following were
purchased from Fisher (Fair Lawn, NJ, USA): hydrochloric acid,
ammonium acetate and glacial acetic acid. The ammonium acetate
buffer used was prepared by weighing 0.77 grams of ammonium acetate
in approximately 700 ml of Milli-Q water. The pH was adjusted to 4,
5.5 or 7 using glacial acetic acid and the v/v methanol/buffer
adjusted accordingly to one liter. After the v/v methanol/buffer was
made to one liter, the mobile phase was then filtered and degassed
before use. The methanol, HPLC grade, was purchased from Burdick and
Jackson Labs (Muskegon, MI, USA). The buffers were prepared using
water from a Millipore (Milford, MA, USA) Milli-Q water system.

RESULTS AND DISCUSSION

A plot of the logarithm of the capacity factor (k') against the
inverse of the absolute temperature (T) for naphthalene and biphenyl
is given in Figure 1. The temperature range used was 25°C to 65°C.
At higher temperatures the retention times obtained were close to the
void volume. Figure 1 shows a linear relationship between k' and
1/T. The slope of the line gives the enthalpy value ($\Delta H°$) for
naphthalene (2.81 KCal/mole) and for biphenyl (3.59 KCal/mole). A
slight decrease was observed in the separation factor α (Table I)
where α is defined as $t_{r'2}/t_{r'1}$ where $t_{r'}$ is corrected retention time
($t_{r1} - t_{ro}$) and t_{ro}, t_{r1} amd t_{r2} are retention times of unretained
peak, naphthalene and biphenyl, respectively.

Table I   Effect of Temperature on α values of naphthalene
and biphenyl

Temperature °C	α
25	2.62
35	2.51
45	2.47
55	2.44
65	2.44

The corrected retention times for naphthalene and biphenyl decreased by 83% each, by changing the column temperature from 25°C to 65°C. This indicates that (a) the interaction of these two polynuclear aromatic hydrocarbons with the beta-cyclodextrin cavity is the same under similar experimental conditions, i.e., mobile phase composition and column temperature, and (b) that the best separation is obtained at the lowest temperature. Figure 2 shows a plot of lnk' vs 1/T for anthraquinone, methyl-and ethylanthraquinone between 25°C and 55°C. At temperatures higher than 55°C the retention times were too close to the void volume to give any meaningful data. The $\Delta H°$ values obtained for anthraquinone (A), methyl- (MA) and ethylanthraquinone (EA), were 2.69, 3.18 and 3.59 KCal/mole respectively. The separation factor values for each pair are given in Table II. Again α values showed a slight decrease with temperature variations. These results are in agreement with those observed by Colin et al. (10) for reversed phase material. The larger the temperature the smaller the α value.

Table II   Effect of temperature on α values of anthraquinones.

Temp °C	α (MA/A)	α (EA/MA)
25	1.53	1.51
35	1.48	1.50
45	1.45	1.48
55	1.44	1.47

The effect of change in column temperature on the retention of anthraquinone, methyl- and ethylanthraquinones between 25°C and 55°C was a decrease of 75%, 76% and 77% respectively.

Figure 3 shows the effect of temperature on the capacity factor of p-nitroaniline, from 0°C to 77°C. A mobile phase consisting of 10% methanol/water was employed. The retention at 0°C was 23.62 min. while at 77°C was 2.28, a ten fold decrease. This decrease in retention may be attributed to many factors such as increased solubility of the p-nitroaniline with increase in temperature, which results in less solute-stationary phase, and an increase in solute-mobile phase interactions; increase in mass transfer, and decrease in the pressure. Also the binding constant of any solute with cyclodextrin goes to zero at 80°C (11).

EFFECT OF METHANOL IN THE ELUENT ON RETENTION. The effect of increasing the volume of the organic modifier, methanol, in the mobile phase on the retention of methyl anthraquinone and naphthalene is given in Figure 4. Methyl anthraquinone and napthalene are more soluble in methanol than in water so increasing the volume of methanol in the mobile phase should result in the increase in the solubility of both compounds, and as a result, a decrease in the retention time. Also, the presence of methanol in the mobile phase affects retention when cyclodextrin bonded columns are used. Methanol is much more tightly bound in the cyclodextrin cavity than

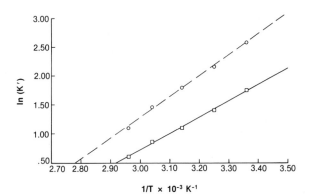

Figure 1.   Effect of temperature on the capacity factor of
naphthalene (-□-) and biphenyl (-0-) using a β-
cyclodextrin column, 4.6 x 100 mm, and a mobile phase
of 45% methanol/water at a flow rate of 1 ml/min.

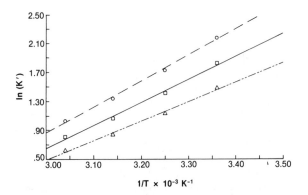

Figure 2.   Effect of temperature on the capacity factor of
anthraquinone (..Δ..), methyl anthraquinone (-□-) and
ethyl anthraquinone (-0-). Experimental conditions
are the same as in Figure 1.

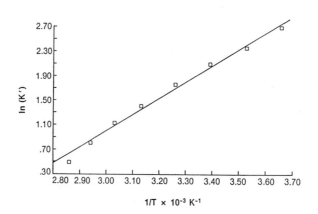

Figure 3.   Same as in Figure 1, but p-nitroaniline and a mobile
phase of 10% methanol/water.

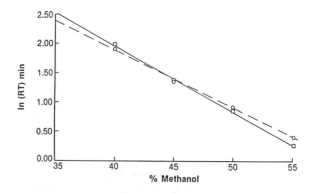

Figure 4.   Effect of volume of methanol in the eluent on the
retention times of methyl anthraquinone (-□-) and
naphthalene (-O-) using a β-cyclodextrin column, 4.6 x
100 mm, and a flow rate of 1 ml/min.

water, but less tightly than most hydrophobic solutes. Since methanol is present at high concentrations compared to the solutes, methanol can displace a solute from the cyclodextrin cavity (11).

EFFECT OF TEMPERATURE AND pH ON RETENTION. It was observed in a previous study (12) that an increase in the temperature resulted in a decrease of retention of a selected group of dipeptides, which is an expected behavior. However, when the pH of the mobile phase was increased from 4.0 (Table III) to 5.5 (Table IV) the retention of some of the dipeptides increased with an increase in temperature.

Table III Effect of column temperature on the retention of dipeptides on a β-cyclodextrin column using a mobile phase of 10% MeOH/0.1 M ammonium acetate, pH 4

Dipeptide	RETENTION TIME (MIN.)						% change t_{r77}/t_{r27}
	27°C	37°C	47°C	57°C	67°C	77°C	
Phe-Gly	3.8	3.7	3.6	3.5	3.5	3.5	-14
Phe-Ala	3.6	3.5	3.5	3.5	3.4	3.4	- 8
Phe-Val	3.9	3.8	3.7	3.7	3.7	3.6	-10
Phe-Met	4.5	4.2	4.1	4.0	3.9	3.9	-20
Phe-Pro	7.1	6.1	5.4	4.8	4.5	4.2	-50
Phe-Ile	4.7	4.4	4.3	4.1	4.0	4.0	-22
Phe-Leu	5.0	4.7	4.5	4.4	4.2	4.1	-25
Phe-Tyr	6.9	5.7	5.2	4.8	4.5	4.3	-47
Phe-Phe	7.3	6.4	5.9	5.5	5.2	5.0	-39
Phe-Asp	8.3	7.8	7.7	7.5	7.2	7.1	-17
Phe-Glu	7.9	7.5	7.4	7.2	7.0	6.9	-16

For example, the retention times of Phe-Ala decreased by 8% by changing the temperature of separation from 27°C to 77°C, at pH 4, but increased by 12% at pH 5.5, and decreased by 9% at pH 7, Table V, using the same experimental conditions. The same was true for Phe-Val and Phe-Ile. At pH 4.0 the retention times of Phe-Asp decreased by 17%, at pH 5.5 it decreased by 15% but increased by 6% at pH 7. In general, however, retention times decreased for the other dipeptides, Tables III-V, with an increase in temperature, but the amount of decrease was varied for different dipeptides. Those with the highest retention times at a certain pH did not decrease the most by changing the separation temperature from 27°C to 77°C. For example, Phe-Pro, has a t_r of 7.1 min. at pH 4 and 27°C, but a t_r of 4.2 min. at 77°C, a decrease of 50%, while Phe-Asp has a t_r of 8.3 min at pH 4 and 27°C, and a t_r of 7.1 at 77°C, a decrease of 17%. Also, the one with the lowest t_r (Phe-Ala, pH 4) did not decrease the most, except at pH 5.5 (Phe-Gly). Therefore, it could be concluded that the degree of decrease in retention with an increase in temperature, is not a function of t_r at 27°C, but of the structure and physical properties of the molecule in question, and the properties of the mobile phase, aqueous, pH, buffer, ionic strength

266   ORDERED MEDIA IN CHEMICAL SEPARATIONS

Table IV   Effect of column temperature on the retention of dipeptides
           on a β- cyclodextrin column using a mobile phase of 10%
           MeOH/0.1 M ammonium acetate, pH 5.5

| Dipeptide | 27°C | R E T E N T I O N   T I M E   (MIN.) | | | | | % change t_{r77}/t_{r27} |
		37°C	47°C	57°C	67°C	77°C	
Phe-Gly	4.2	4.2	4.1	4.1	4.1	4.2	0
Phe-Ala	4.2	4.2	4.2	4.2	4.3	4.5	+12
Phe-Val	4.7	4.8	4.9	5.0	5.2	5.3	+17
Phe-Met	5.4	5.3	5.2	5.2	5.3	5.3	- 1
Phe-Pro	7.8	6.9	6.1	5.6	5.3	5.0	-43
Phe-Ile	6.0	6.0	5.9	6.0	6.1	6.1	+ 2
Phe-Leu	6.3	6.2	6.1	6.1	6.1	6.0	- 5
Phe-Tyr	8.0	7.8	7.2	6.8	6.5	6.3	-35
Phe-Phe	9.2	8.4	7.7	7.6	7.4	7.2	-25
Phe-Asp	8.3	7.9	7.7	7.5	7.4	7.2	-15
Phe-Glu	8.7	8.3	8.1	7.9	7.8	7.8	-12

Table V   Effect of column temperature on the retention of dipeptides
          on a β-cyclodextrin column using a mobile phase of 10%
          MeOH/0.1 M ammonium acetate, pH 7

| Dipeptide | 27°C | R E T E N T I O N   T I M E   (MIN.) | | | | | % change t_{r77}/t_{r27} |
		37°C	47°C	57°C	67°C	77°C	
Phe-Gly	4.6	4.6	4.5	4.5	4.4	4.2	-13
Phe-Ala	4.7	4.8	4.8	4.7	4.5	4.4	- 9
Phe-Val	5.5	5.6	5.5	5.3	5.1	4.9	-16
Phe-Met	6.3	5.8	5.6	5.3	5.0	4.8	-30
Phe-Pro	8.1	7.0	6.3	5.7	5.2	4.9	-48
Phe-Ile	7.1	6.6	6.3	5.9	5.7	5.4	-30
Phe-Leu	7.3	6.7	6.3	5.9	5.7	5.4	-33
Phe-Tyr	10.6	8.4	7.3	6.5	6.1	5.5	-55
Phe-Phe	11.9	10.0	8.9	8.2	7.7	6.9	-48
Phe-Asp	4.3	4.3	4.4	4.1	4.3	4.5	+ 6
Phe-Glu	4.5	4.6	4.6	4.4	4.7	4.8	-11

and organic modifier. It should be emphasized that the mobile phase
used in this study was 5% methanol in 95% 0.005 M ammonium acetate,
pH 4, 5.5 or 7. Horvath and coworkers (13) in studying the role of
acidic amine phosphate buffers as eluents observed that "besides
their classical static role to maintain the pH of a solution
constant, buffers may play a variety of other roles and affect
significantly the properties and efficiency of a chromatographic
system". One of these effects is the masking of the silinol groups
at the stationary phase surface by the weak amine component of the
buffer.

TEMPERATURE AS A SEPARATION PARAMETER. While temperature may be used to improve the separation factor, a quick glance at Tables III-V, reveal that in most cases the above statement is not true, but in certain cases it holds. Phe-Phe and Phe-Glu had retention times of 7.3 and 7.9 at 27°C respectively, while the retention times at 77°C are 5.0 and 6.9 which is an increase in $t_{r2} - t_{r1}$ from 0.6 to 1.9 min. Other examples can be found in Tables III-V. Such temperature effects are not unusual. Chemielowiec and Sawatzky (14) used temperature as a separation parameter for a group of polynuclear aromatic hydrocarbons (PAH) on C_{18} reversed phase column using a mobile phase of acetonitrile/water. They observed that the elution order of some PAH are reversed by temperature variations, which are "entropy dominated separations". Snyder (15) in commenting on the above study (14) and the effect of temperature on retention concluded that in a chromatographic system the dependence of the capacity factor k' on temperature is determined by the enthalpy (ΔH°):

$$d(\log k')/d(1/T) = \Delta H°/4.57 \qquad (1)$$

Where T is the absolute temperature and ΔH° is in calories.

According to Snyder (15), "if the linear dependence of ΔH° on log k' is exact, then solute retention order will be unchanged as separation temperature is varied". This is referred to as "regular" temperature behavior. He concludes that "temperature is not an effective parameter for altering α (the separation factor) values and maximizing resolution in "regular" systems. However, failure of relationships such as equation 1 may be due to: (1) retention of solute molecules by more than one mechanism or (2) marked difference in the molecular shapes of two solutes with similar retention in a particular LC system. It is worth while to examine the results in light of Horvaths (13) and Snyders (15) observations, remembering that we are dealing with a buffered mobile phase and a beta cyclodextrin bonded silica column.

A plot of lnk' vs 1/T at each pH should reveal the effect of pH and T on the properties and efficiency of the chromatographic system. Figure 5 shows a plot of lnk' vs 1/T at each of the mobile phase pH's studied. It is clear that a linear relationship exists at pH 4 and pH 5.5 which has similar slopes (ΔH°). The linear relationship indicates that the same mechanism of separation is taking place throughout the experiment 27°C - 77°C. However, at pH 7 it is clear that there exists two different mechanisms of separation. The same is true for Phe-Asp, Figure 6. The noticeable difference between Figures 5 and 6 is the effect of pH on retention at pH 4 and 5.5. Plots of lnk' vs 1/T for Phe-Gly (Figure 7) and Phe-Ile (Figure 8) show that one mechanism is involved at pH 7 while two different mechanisms are involved at pH 4 and 5.5. For Phe-Tyr, the linear relationship (one mechanism) is observed at pH 5.5 while two mechanisms of separation are each observed at pH 4 and pH 7, Figure 9. Different results from the above (Figures 4-8) were observed for plot of lnk' vs 1/T for Phe-Val. At pH 5.5 the value of lnK' decreased, Figure 10, while in the other cases (Figures 4-8) it increased. Also, note that although two different mechanisms are

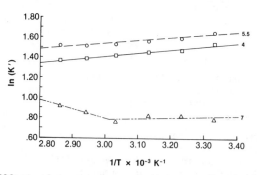

Figure 5. Effect of temperature on the capacity factor of Phe-
Glu using a β-cyclodextrin column and a mobile phase of
5% methanol/0.005 M ammonium acetate at a pH of 4
(-□-), 5.5 (-O-) and 7 (..Δ..) at a flow rate of 1
ml/min.

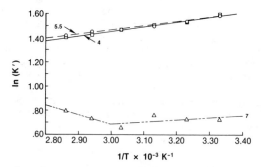

Figure 6. Same conditions as in Figure 5, but Phe-Asp.

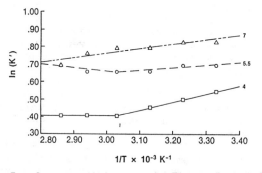

Figure 7. Same conditions as in Figure 5, but Phe-Gly.

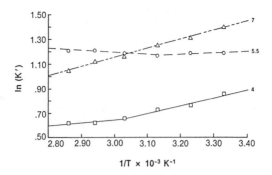

Figure 8. Same conditions as in Figure 5, but Phe-Ile.

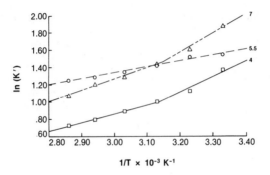

Figure 9. Same conditions as in Figure 5, but Phe-Tyr.

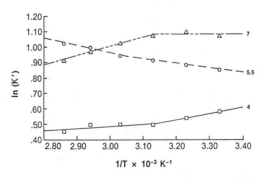

Figure 10. Same conditions as in Figure 5, but Phe-Val.

involved at each, pH 4 and pH 7, the slopes are different and opposite of each other. The above examples, Figures 5-10, clearly indicate the role a buffer may play and how it effects the properties and efficiency of the chromatographic system. These results, therefore, agree with what was observed earlier by Horvath et al. (13) and others (14,15). Another explanation for the non linear relationship between lnk' vs 1/T that may be postulated here is that each inclusion complex will dissociate at a distinct temperature. It is possible that the breaks in the curves indicate this temperature (11).

ACKNOWLEDGMENTS

The authors would like to thank Dr. D. Armstrong (Texas Tech. University) for helpful discussion and constructive comments. By acceptance of this article, the publisher or recipient acknowledges the right of the U.S. Government to retain a nonexclusive, royalty-free license and to any copyright covering the article. This project has been funded at least in part with Federal funds from the Department of Health and Human Services, under contract number N01-CO-23910 with Program Resources, Inc. The contents of this publication do not necessarily reflect the views of the Department of Health and Human Services, nor does mention of trade names, commercial products, or organizations imply endorsement by the U.S. Government.

LITERATURE CITED

1. Armstrong, D.W., Alak, A., DeMond, W., Hinze, W.L. and Riehl, T.E., J. Liquid Chromatogr. 1985, 18, 261-269 and references there in.

2. Armstrong, D.W. and DeMond, W., J. Chromatogr. Sc. 1984, 22, 411-415.

3. Armstrong, D.W., DeMond, W., and Czech, B.P., Anal. Chem. 1985, 57, 481-484.

4. Ward, T.J. and Armstrong, D.W., J. Liq. Chromatogr. 1986, 9, 407-423 and references there in.

5. Abidi, S.L., J. Chromatogr. 1986, 362, 33-46.

6. Issaq, H.J., Weiss, D.E., Ridlon, C., Fox, S.D. and Muschik, G.M., J. Liq. Chromatogr. 1986, 9, 1791-1801.

7. Issaq, H.J., McConnell, J.H., Weiss, D.E., Williams, D.G. and Saavedra, J.E., J. Liq. Chromatogr. 1986, 9, 1783-1790.

8. Issaq, H.J., J. Liq. Chromatogr. 1986, 9 229-233.

9.   Hinze, W.L., Riehl, T.E., Armstrong, D.W., DeMond, W., Alak, A., Ward, T., Anal. Chem. 1985, 57, 237-242.

10. Colin, H., Diez-Masa, J.C., Guiochon, G., Czajkowska, T. and Miedziak,
     I., J. Chromatogr. 1978, 167. 41-65.

11. Armstrong, D.W., Private communication.

12. Ridlon, C. and Issaq, H.J., J. Liq. Chromatogr., in press.

13. Melander, W.R., Stoveren, J. and Horvath, Cs., J. Chromatogr. 1979, 185, 111-127.

14. Chemielowiec, J. and Sawatzky, H., J. Chromatogr. Sci. 1979, 17, 245-252.

15. Snyder, L.R., J. Chromatogr. 1979, 179, 167-172.

RECEIVED May 11, 1987

Chapter 16

Computer Imaging of Cyclodextrin Inclusion Complexes

R. Douglas Armstrong

La Jolla Cancer Research Foundation, 10901 North Torrey Pines Road, La Jolla, CA 92037

X-ray crystal structures were used for the production of computer projected images of inclusion complexes of structural isomers, enantiomers and diastereomers with α- or β-cyclodextrin. These projections allow for a visual evaluation of the interaction that occurs between various molecules and cyclodextrin, and an understanding of the mechanism for chromatographic resolution of these agents with bonded phase chromatography.

The wide interest in the use of cyclodextrins as a separation medium has led to a number of useful applications. The ability of these molecules to bind other molecules to form an inclusion complex, has provided for their use in typically difficult separations of enantiomers, diasastereomers, and structural isomers. Through the coupling of cyclodextrin to a solid support, such as silica gel, a chromatographic resin can be made, and has been developed as a useful chromatographic procedure.

Although it is well understood that molecules must be able to enter the cavity of the cyclodextrin molecule for complexation to occur, and therefore, under chromatographic conditions, for retention to result, the differential binding of two stereoisomers within the cyclodextrin that allows for their differential retention is not always apparent. An understanding of this can be obtained through the use of three dimensional computer graphic imaging of the crystal structure of the inclusion complex.

This review will illustrate examples of computer projected models of inclusion complexes of structural isomers (ortho, meta, para nitrophenol), enantiomers (d- and l- propranolol) and diastereomers [cis and trans 1(p-β-dimethylaminoethoxy-phenyl-butene), tamoxifen] in either α- or β-cyclodextrin. The use of these computer projections of the crystal structures of these complexes allows for the demonstration and prediction of the chromatographic behavior of these agents on immobilized cyclodextrin.

0097-6156/87/0342-0272$06.00/0

Computer Projected Inclusion Complexes

Structural Isomers

Cyclodextrin bonded phases have been demonstrated to be particularly adept in resolving structural isomers (1) such as the ortho, meta and para forms of nitrophenol (Figure 1). These molecules can be separated on columns of either β or α-cyclodextrin, the respective binding constants (at pH 10) for α-CD inclusion complexation are 200, 500, and 2439 M^{-1} respectively for ortho, meta and para nitrophenol (1). Figure 2 illustrates the computer projection of each of these complexes based on their x-ray crystal structures (2,3,4). The pictures illustrate a side view of the complex with the front of the cyclodextrin cut away in order to see the degree of penetration of the complexed molecule. The para-nitrophenol (red molecule in Figure 2a), is found to complex deep in the cyclodextrin cavity, with the nitro-portion of the molecule in potential position to interact with the lower 6-hydroxyl atoms of the α-cyclodextrin. The deep and centered penetration of para-nitrophenol also allows for excellent interaction of the phenol ring with the nonpolar cyclodextrin cavity. This is in contrast to the meta-nitrophenol (green molecule, Figure 2b), which also significantly enters the cyclodextrin cavity, but to a lesser degree than does the para-nitrophenol. This allows for a reduced interaction with the α-cyclodextrin, and the lower binding constant (500 compared to 2439 M^{-1} for the para-nitrophenol). The ortho-nitrophenol (yellow molecule, Figure 2c), which exhibits the lowest binding constant with α-cyclodextrin at 200 M^{-1}, is clearly illustrated to have the least penetration and complexation in the α-cyclodextrin cavity. The location of the nitro group when in the ortho position blocks the molecule from entering the cyclodextrin cavity, as observed from the closeness of the van der Waals' radii. The computer projected inclusion complexes nicely demonstrate the reasons for the variable binding constants of these structural isomers, and are very predictive of their resulting chromatographic behavior.

There have been several reports which have demonstrated the chromatographic separation of ortho, meta and para structural isomers on β-cyclodextrin matrices (5). One common feature of these separations is that the order of retention (greatest to lowest) falls in the order of para > ortho > meta, although the ortho and meta isomers elute very close to one another. This is in contrast to what is obtained, as illustrated above for nitrophenol, with α-cyclodextrin, where the binding affinity for the meta isomer is greater than the ortho. In an attempt to understand this anomaly, the complex of ortho or meta-nitrophenol in β-cyclodextrin was modeled. As expected from the much greater capacity of the β-cyclodextrin (composed of 7 glucose units) compared to the alpha-cyclodextrin (composed of 6 glucose units), the small nitrophenol molecules both easily fit into the β-cyclodextrin. The blocked entry problem that the ortho-nitrophenol exhibited in the α-cyclodextrin, does not exist in the β-cyclodextrin, and allows for even greater complexation than what is observed for meta-nitrophenol (binding constants of 357 M^{-1} for

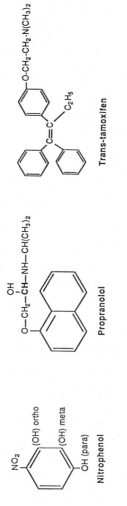
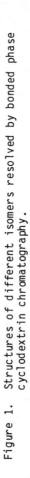

Nitrophenol

Propranolol

Trans-tamoxifen

Figure 1. Structures of different isomers resolved by bonded phase cyclodextrin chromatography.

ortho-nitrophenol compared to 147 M^{-1} for meta-nitrophenol, ref. 1).

Enantiomers

Potentially, one of the most valuable applications of cyclodextrin as an analytical tool is its use in resolving enantiomeric compounds, those compounds which are mirror images of each other. This is an important concern with synthetic pharmaceuticals, which are often produced as enantiomers. In most cases, both isomers can have physiological activity, although only one actually has capacity to produce the desired therapeutic action. The inactive isomer will often contribute to host toxicity or other undesired actions which can limit the effectiveness of the active isomer. The ability of cyclodextrin to resolve many types of enantiomers is of obvious benefit, and has been demonstrated for a number of relevant pharmaceuticals (6).

One important drug that is synthesized as both d and l enantiomers, which can be resolved using immobilized β-cyclodextrin, is propranolol (Figure 1). Under standard chromatographic conditions (see ref. 6), the d-propranolol is retained much longer than is the l-propranolol. The respective inclusion complex of each isomer in β-cyclodextrin is illustrated in the computer projections in Figure 3 a and b. This illustrates that there is no difference between d and l propranolol in their actual complexation within the β-cyclodextrin cavity, as the napthol rings of each compound assume the exact same placement within the β-cyclodextrin. However, very important differences exist from the point of the chiral carbon on the aliphatic side chain. In contrast to the ortho, meta, para structural isomers, which were shown to resolve because of their respective abilities to be complexed within the cyclodextrin cavity, for enantiomeric resolution, the unidirectional 2- and 3-hydroxyl groups located at the mouth of the cyclodextrin cavity appear to be integral for chiral recognition. In the models illustrated in Figure 3, the van der Waals' radii are shown for only these 2- and 3-hydroxyl groups of the β-cyclodextrin, along with the secondary amine of the propranolol molecules. The hydroxyl group attached to the chiral carbon of propranolol is in the same position for the d and l isomers, and is placed for optimal hydrogen bonding to a 3-hydroxyl of the cyclodextrin. Important differences are observed, however, between the d and l forms with respect to their secondary amine group. In the d-propranolol complex, the nitrogen is ideally situated for hydrogen bonding with both a 2- and 3-hydroxyl group on the β-cyclodextrin, exhibiting bond distances of 3.3 and 2.8Å. The amine of the l-propranolol however, is less favorably situated for hydrogen bonding, with distances of 3.8 and 4.5Å to the closest 2- and 3-hydroxyl groups of the β-cyclodextrin which are too great for hydrogen bonding. The gap between the van der Walls' radii of these atoms is clearly seen in the l- propranolol model (Figure 3b), whereas the van der Waals' radii in the d- propranolol are very closely associated with those for the 2- and 3-hydroxyl groups on the β-cyclodextrin (Figure 4a). With the ability to form additional hydrogen bonds, the d-propranolol exhibits a stronger binding with the β-cyclodextrin, and is thereby retained longer under bonded phase chromatographic conditions (6). Therefore, two parameters are important for chiral resolution: 1. the ability of the compound to form an inclusion complex within the cyclodextrin cavity, which provides for retention of the compound, and 2. the interaction of the portion of the molecule containing the chiral carbon with the unidirectional 2- and 3-hydroxyl groups on the cyclodextrin.

Diastereomers

The ability of cyclodextrin to resolve stereoisomers is very readily applied for the separation of diastereomers, such as the cis- and trans-geometric isomers (6,7). Similar to both of the examples presented above, resolution of geometric isomers appears to result from both the level of inclusion complex formed, as well as the level of interaction of the molecule with the 2- and 3-hydroxyl groups of the cyclodextrin. This can be illustrated with the synthetic antiestrogen tamoxifen (Figure 1), which is synthesized in both the cis and trans forms.

These two compounds can be separated using β-cyclodextrin bonded phase chromatography, with the cis-tamoxifen eluting prior to the trans-tamoxifen (6,7). Using the individual x-ray crystal structures for these agents (8,9), the respective inclusion complexes in β-cyclodextrin were modeled using computer imaging, as illustrated in Figure 4, a (trans-tamoxifen) and b (cis-tamoxifen). It is very apparent that these two agents interact quite differently with the β-cyclodextrin. The trans-tamoxifen is able to form a better inclusion complex than can the cis-tamoxifen, with its phenylside group penetrating 6.3Å (measured from the 3-hydroxyl of the β-cyclodextrin to the lowest atom of the respective molecule of tamoxifen) compared to the 5.7Å penetration of the cis-tamoxifen. In addition to the greater level of complexation, it appears that the trans-tamoxifen may have some additional interaction between its aliphatic side-chain and the mouth of the β-cyclodextrin.

Summary

The usefulness of cyclodextrin as a separation medium for the resolution of steroisomers, whether they be structural isomers, diastereomers or enantiomers, has become readily apparent. The understanding of how and why a particular separation occurs is nicely enhanced with the use of computer modeling of the x-ray crystal structures of the agents. For example, the computer modeling of the compounds in this review illustrated the importance of the 2- and 3- hydroxyl groups for the resolution of enantiomers, whereas differential inclusion complexation was demonstrated to mediate the resolution of structural isomers such as ortho, meta and para nitrophenol. In particular, although not shown in the above results, the use of the computer imaging may greatly improve efforts to rationally derivatize cyclodextrin in order to optimize a particular separation. It has been demonstrated in other reports that cyclodextrin is limited in its capability to serve as an effective medium for the resolution of enantiomeric compounds, and chiral compounds that form very good inclusion complexes are not

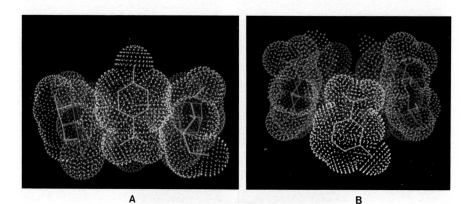

A

B

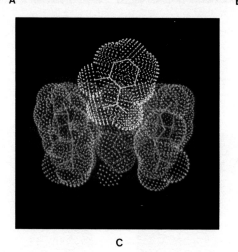

C

Figure 2.   Computer imaging of crystal structures of the inclusion complexes of para (A), meta (B) and ortho (C) nitrophenol with α-cyclodextrin. The complex is shown with van der Waals' radii, and the front section of the complex cut away in order to expose the nitrophenol molecule.

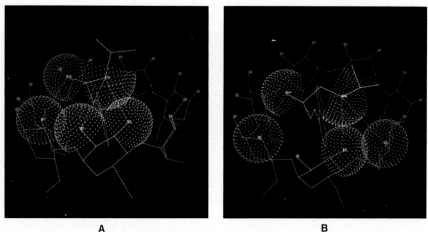

A   B

Figure 3.   Computer imaging of the inclusion complexes of d-(A) and
            l-(B) propranolol with β-cyclodextrin.   The chemical
            structures are illustrated with van der Waals' radii
            shown for only the secondary amine of propranolol and
            the 2- and 3- hydroxyl groups of the β-cyclodextrin.

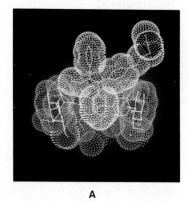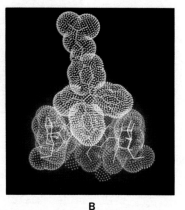

A   B

Figure 4.   Computer imaging of the inclusion complexes of trans-
            (A) and cis- (B) tamoxifen with β-cyclodextrin.   The
            complex is shown with van der Waals' radii, and the
            front section of the complex cut away in order to expose
            the nitrophenol molecule.

always necessarily resolved from each other (6). In some cases however, a small derivatization modification of the cyclodestrin can allow for the needed separation (10). With improved and easier methods of computer modeling and energy minimization calculations, combined with the lowering cost of obtaining such a system, the use of computer imaging should continue to be a most valuable resource in the study of cyclodextrins and their varied functions.

Acknowledgments

This work was supported by grant CH 329 from the American Cancer Society. The computer imaging was completed at the Computer Graphics Laboratory (Dr. R. Langridge, director; supported by NIH grant RR 1081) with the assistance of N. Pattabiraman.

References

1. Hinze, W.. Separation and Purification Methods, 1981. 10(2):159-237.

2. Harata, K., Hisahi,.U. and Tanaka, J. Bull. Chem. Soc. Japan, 1978. 51:1627-1634.

3. Harata, K. Bull. Chem. Soc. Japan, 1977. 51;1416-1424.

4. Iwqsaki, F. and Kawano, Y. Acta. Cryst., 1978. 34:1286-1290.

5. Armstrong, D. W., DeMond, W., Alak, A., Hinze, W. L., Riehl, T. E. and Bui, K. H. Analy. Chem. 1985. 57(1):234-237.

6. Armstrong, D. W., Ward, T. J.,.Armstrong, R. D. and Beesley, T. E. Science. 1986, 232:1132-1135.

7. Armstrong, R. D., Ward, T. J., Pattabiraman, N., Benz, C. and Armstrong, D. W. J. Chromatog 1987. 414:192-196.

8. Kilbourn, B. T. and Owston, P. G. J. Am. Chem. Soc. 1970. 736:1-5.

9. Precigoux, P. G., Courseille, C., Geoffre, S. and Hospital, M. Int. Union Cryst. 1979. 12:3070-3072.

10. Armstrong, D. W., Ward, T. J., Czech, A., Czech, R. A. and Bartsch, R. A. J. Org. Chem. 1985. 50:5556-5559.

RECEIVED February 17, 1987

INDEXES

Author Index

Affiliation Index

Subject Index

Production by Barbara J. Libengood
Indexing by Keith B. Belton
Jacket design by Carla L. Clemens

Elements typeset by Hot Type Ltd., Washington, DC
Printed and bound by Maple Press Co., York, PA

Recent ACS Books

Personal Computers for Scientists: A Byte at a Time
By Glenn I. Ouchi
288 pp; clothbound; ISBN 0–8412–1001–2

The ACS Style Guide: A Manual for Authors and Editors
Edited by Janet S. Dodd
264 pp; clothbound; ISBN 0–8412–0917–0

Silent Spring Revisited
Edited by Gino J. Marco, Robert M. Hollingworth, and William Durham
214 pp; clothbound; ISBN 0–8412–0980–4

Chemical Demonstrations: A Sourcebook for Teachers
By Lee R. Summerlin and James L. Ealy, Jr.
192 pp; spiral bound; ISBN 0–8412–0923–5

Phosphorus Chemistry in Everyday Living, Second Edition
By Arthur D. F. Toy and Edward N. Walsh
342 pp; clothbound; ISBN 0–8412–1002–0

Pharmacokinetics: Processes and Mathematics
By Peter G. Welling
ACS Monograph 185; 290 pp; ISBN 0–8412–0967–7

Solving Hazardous Waste Problems
Edited by Jurgen H. Exner
ACS Symposium Series 338; 397 pp; ISBN 0–8412–1025–X

Crystallographically Ordered Polymers
Edited by Daniel J. Sandman
ACS Symposium Series 337; 287 pp; ISBN 0–8412–1023–3

Pesticides: Minimizing the Risks
Edited by Nancy N. Ragsdale and Ronald J. Kuhr
ACS Symposium Series 336; 183 pp; ISBN 0–8412–1022–5

Nucleophilicity
Edited by J. Milton Harris and Samuel P. McManus
Advances in Chemistry Series 215; 494 pp; ISBN 0–8412–0952–9

Organic Pollutants in Water
Edited by I. H. Suffet and Murugan Malaiyandi
Advances in Chemistry Series 214; 796 pp; ISBN 0–8412–0951–0

For further information and a free catalog of ACS books, contact:
American Chemical Society
Distribution Office, Department 225
1155 16th Street, NW, Washington, DC 20036
Telephone 800-227-5558